Jimmy **Teng**

Musket, Map and Money:
How Military Technology Shaped Geopolitics and Economics

VERSITA

Versita Discipline: Economics

Managing Editor
Ewa Feder-Sempach

Language Editor
Thomas E Dudley

Published by Versita, Versita Ltd, 78 York Street, London W1H 1DP, Great Britain.

ISBN (paperback): 978-83-7656-057-1

ISBN (hardcover): 978-83-7656-058-8

ISBN (for electronic copy): 978-83-7656-059-5

Managing Editor: Ewa Feder-Sempach

Associate Editor: Urszula Mrzygłód-Opacka

Language Editor: Thomas E Dudley

www.versita.com

Cover illustration: © istockphoto.com/mikdam

Contents

CHAPTER 1

Introduction

How did we get to the modern western centered world? Modernization is synonymous with westernization, and western nations dominating current world affairs, the western economies and companies generate most of the world's nominal wealth with western ideas and values shaping the world's cultures. To answer that is to address what caused the rise of the west above all other major cultures of the world, including Islamic, Indian, Chinese and Japanese, in the early modern era. To view it in a broader perspective, what are the forces that shaped the general contours of world history and the rise and fall of civilizations?

The inquiry into the accumulation and distribution of wealth amongst nations is the raison d'etre of economics.[1] But even now, this inquiry has not yet been completed with many conundrums still exist. One conundrum that exists in economics development and economic history in particular, is of course the spectacular economic rise of the West. The economic rise of the West, also termed the European miracle within the literature, refers to the industrialization and modernization of Europe ahead of all the other Eurasian major cultures in the early modern era.[2] Growth rates of European economies were higher than the world average throughout several centuries starting from circa 1400 AD. From ca 1400 AD onwards, Europe underwent those economic, political, social, technological and geographical upheavals which were to make it the birthplace of the industrial world. The advance of Europe relative to the rest of the world was in all spheres of human endeavors: Europe forged ahead economically, politically and militarily. The industrial revolution of England and then Europe was part of this phenomenon. This is near-miraculous, for during

---

1 Refer to Smith (1776).
2 Refer to Pomeranz (2000) and Jones (2002).

the medieval era, the Arabian and the Chinese world (which was the Cathay of Marco Polo), as well as the Indian subcontinent, were ahead of Europe in terms of economic development.

The rise of Europe might seem unlikely initially, for from ca 1200 AD onwards, only Europe operated under a competitive state system, with rivals (almost evenly-matched in terms of military and economic might) constantly competing for power. Other civilizations were ruled by continental-size empires. Europe suffered from constant interstate wars and differences in languages and administrative practices. On the other hand, the other civilizations enjoyed peace, uniformity in administrative practices and the convenience of an official language used across a unified empire. Nevertheless, despite these disadvantages, Europe modernized while the rest of the world stagnated. This energy of Europe and relative inertia of the rest of the world was maintained for centuries until the gap was so wide that European global supremacy was established: Europe entered the modern world, while the other major cultures had remained medieval or even regressed. The big question is: where did the sudden spurt of energy in the European economies come from?[3]

The difficulty in explaining the European miracle is not unique: there are many other instances of very long-term economic changes that are also hard to account for. A well-known case is the high economic achievement of China during the Song dynasty. The high economic achievement of the Song Dynasty is called the Song puzzle, for this high achievement was not repeated in the later history of China.[4] The Chinese economy stagnated or even declined in the later centuries. Other similar puzzles include the leadership of Mesopotamia in early civilizations, which failed to continue during the classical era; the explosions in human achievements in Greece, Ganges India and China during the axial age; and, the Eurasian dominance over America, Africa and Oceania in the development of civilization.

What caused these instances of very long term (that is, a few centuries or longer) economic progress and decline? The inquiry into very long term economic performance is a separate field by itself, different from most economic studies that focus on the medium or short term performances.[5] Most economic enquiries have their explanations at the level of individual economic agents, that is, the consumers or the firms. This is named micro foundation, the hallmark of mainstream neoclassical economics. North (1987, 1990) argues that these theories explain the mechanism of growth but not the causes of growth.

---

3 This is the question raised by Kennedy (1987) at the beginning of the book.
4 Refer to Elvin (1973) and Jones (1981, 1988, 1990).
5 Refer to Schumpeter (1911), Baran (1957), Franks (1975), Martinussen (1997) and Barro and Sala-i-Martin (1998) for related treatments.

North (1987, 1990) posits that factors that are important in affecting very long term economic performance are different from those that are critical in determining medium or short term performance, which most economists focus on.

The Rise of the Western World: A New Economic History (North and Thomas, 1973) is one of the earliest and most prominent works explaining very long term economic performance. North and Thomas (1973) argue that efficient property rights regimes were the basis of the European miracle.[6] Fluctuations in population in Europe and changes in relative factor prices produced the efficient property rights regimes and institutions that in turn formed the basis for economic progress and the subsequent industrial revolution. The main driving force of progress was the increasing population to land ratio in Europe after the recovery from the scourge of the Black Death. An increasing population to land ratio in Europe generated changes in relative prices and produced efficient property rights regimes and institutions, which as stated earlier, were the basis for the European miracle. North (1979, 1981, 1984) extends the thesis of North and Thomas (1973) and argues that the predatory state which extract resources from economy without institutional check and balances is the main cause of man-made economic declines in history. Levy (1981, 1988), Olson (1993, 2000) and Acemoglu and Robinson (2012) agree along the same line and further the research. Olson (1993) observes that:

> "Individual rights to property and contract enforcement were probably more secure in Britain after 1689 than anywhere elsewhere, and it was in Britain, not very long after the Glorious Revolution, that the Industrial Revolution began." (574)

Another important line of research about the rise of the West studies the role of international political and military rivalry in affecting economic performance. One of the ways that international rivalry could affect economic performance is through influencing the choice and design of political, economic and property rights institutions. Jones (1981) argues that the unique geopolitical environment and state system of Europe was the cause of the European miracle. Competition between states generates concern for national power, including economic and military progression. Consequently, European states tried to outperform each other in all spheres of endeavors that have significance for national power. Europe was propelled forward with great speed due to such international political military competitions.

---

6 Refer to Coase (1937, 1960, 1988) and Barzel (1989) for property rights economics.

North (1981, 1995, 1998) shared the view of Jones (1981). While stressing the role of institutions and property rights regimes in affecting economic performance, North (1981) posits that the competition between European states was a source of institutional change that led ultimately to the European Miracle. Intense and prolonged interstate rivalry led to changes in political and economic institutions as states became more inclusive in order to get support from wider social groups to sustain their effort for greater power at the international arena and that led to better economic performance.[7] North (1995, 1998) reiterates his position in North (1981) to explain economic development by international political military competition. North (1995) notes that:

> "......Even the relative failures in Western Europe played an essential role in European development and were more successful than China or Islam because of competitive pressures." (26)

The "relative failures" here refers to countries such as Portugal or Spain that were once forerunners in European economic development but somehow were overtaken later by countries such as Netherlands and England.

Jones (1988) further develops the basic theme of Jones (1981) in the case studies of Song China and Tokugawa Japan, and Jones (1990) provides a more detailed study of the Song China's high economic achievements. Jones (1988) calls for case studies of other major instances of very long-term economic changes in a world historical framework. Jones (2002) repeats this exhortation.[8]

The call did not go unanswered. Bernholz et al. (1998) and Bernholz and Vaubel (2004) answer the call by formulating the Hume-Kant hypothesis and testing it against practically the whole of world history. Bernholz et al. (1998) and Bernholz and Vaubel (2004) termed the theory that explains economic development by the nature of the international political system the Hume-Kant Hypothesis and presented case studies that cover almost the whole written history of mankind. According to Bernholz and Vaubel, they themselves weren't the originators of the Hume-Kant hypothesis—they see its assertion, under varying names, in many writers through history. The Hume-Kant hypothesis has echoes in many prominent thinkers and scholars from all academic fields over the centuries. Among these thinkers and scholars are Gibbon (1787), Weber (1923), Wesson (1967, 1978), Rostow (1974), Baechler (1976, 1988), Kennedy (1987), Parker (1996) and of course, North (1981, 1995, 1998) and Jones (1981, 1988).[9]

---

7 Lake (1992) observes that democracries have been about twice as likely to win wars as have dictatorships.
8 Refer to Bentley and Adas (1995) and Buzan and Little (2000) for similar endeavors.
9 Wesson (1978, p. 250) cited: "...... As Francis Bacon opined, "No body can be healthy

The Hume-Kant hypothesis argues that the state system, with its pluralistic international power structure, has superior economic performance to the imperial order, where power is monopolized by the imperial regime. David Hume and Immanuel Kant were the earliest advocates of the theory that the state system was the basis of progress of civilization.[10] The following quotations from David Hume and Immanuel Kant give the hypothesis its name:

David Hume (1742, *Of the Rise and Progress of the Arts and Sciences*):

"That it is impossible for the arts and sciences to arise, at first, among any people unless that people enjoy the blessing of a free government." (61)

"That nothing is more favourable to the rise of politeness and learning than a number of neighbouring and independent states, connected together by commerce and policy." (64).

"Where a number of neighbouring states have great intercourse of arts and commerce, their mutual jealousy keeps them from receiving too lightly the law from each other, in matters of taste and of reasoning, and makes them examine every work of art with the greatest care and accuracy." (65)(as cited in Haakonssen eds. (1994, pp. 58-77).)

Immanuel Kant (1784):

"Now the states are already in the present day involved in such close relations with each other that none of them can pause or slacken in its internal civilization without losing power and influence in relation to the rest ... Civil liberty cannot now be easily assailed without inflicting such damage as will be felt in all trades and industries, and especially in commerce; and this would entail a diminution of the powers of the State in external relations ... And thus it is that, notwithstanding the intrusion of many a delusion and caprice, the spirit of enlightenment gradually arises a great good which the human race must derive even from the selfish purposes of aggrandizement on the part of its rulers, if they understand what is for their own advantage." (as cited in Gardiner ed. (1959, pp. 22-34).)

Bernholz (1998) paraphrases the hypothesis as follows:

---

without exercise, neither natural body nor politic, and certainly to a kingdom or estate, a just and honorable war is the true exercise ... for in a slothful peace, both courages effeminate and manners corrupt"*Essays, "Of the True Greatness of Kingdoms"*.
10 Refer to Bernholz and Vaubel (2004, p. 1).

"... military and international political competition among states has forced rulers to grant safe property rights, rule of law and reasonably low and calculable taxes to their subjects. For states following such policies were, in the long run, more successful in this international political competition since they could employ larger resources and were more innovative militarily. But both resources and innovativeness were dependent on favorable economic development. The economy, however, flourished best in states which, by chance or design, introduced safe property rights, a reliable legal system, free markets, stable money, etc. Moreover, citizens well-satisfied with their economic plight and accepting the political regime were presumably better prepared to fight for the survival or even expansion of their country." (109-110)

In testing the Hume-Kant hypothesis, Bernholz et al. (1998) and Bernholz and Vaubel (2004) have case studies that cover almost the whole of world history: ancient Sumer and Phoenicia, classical Greece, the Roman Empire, medieval and modern Europe, imperial China, pre-modern Japan, India and the Islamic world. North (1998) is one of the contributors in Bernholz et al. (1998) and agrees with the Hume-Kant hypothesis. North (1998) posits that:

"The ubiquitous competition among the evolving nation states was a deep underlying source of change and equally a constraint on the options available to rulers within states. It was the competition that forces the Crown to trade rights and privileges for revenue, including, most fundamentally, the granting to "representative" bodies – variously Parliament, Estates General, Cortes – control over tax rates and/or certain privileges in return for revenue. Equally, competition amongst states offered constituents alternatives – states to which they might flee or send their moveable wealth, thus constraining the ruler's options." (24)

Among those that agree with the Hume-Kant hypothesis are Weiss and Hobson (1995). Weiss and Hobson (1995) argue that an important drive behind the American and English as well as other nations' industrialization was the intervention of the state on the economy due to the expectation of coming war, the preparation for war, the actual conduct of war, and the post-war reorganization.[11] Weiss and Hobson (1995) explain that an important cause behind conscious efforts of the state to build a strong economy is the intense military contests that can occur between states. In such contests, economic might and

---

11 Refer to Weiss and Hobson (1995, Ch. 3). Refer to Evans et al. (1985) for the need to bring the state back in when doing social sciences analysis.

industrial capacity are the keys to victory.[12] States with weak militaries and economies suffer defeats and at times are taken over. Weiss and Hobson (1995) point out that with intense military contests, the state has an incentive to correct market and institutional failures and promote economic growth. This explains the early modern European and pre WWII Japanese industrialization and the recent East Asian newly industrializing economies' economic development.

Explaining economic development by international political military competition allows the examination of how natural selection replaces—or doesn't replace—inefficient institutions with efficient ones, and weak states with strong ones. This natural selection mechanism comes from the competition between states through wars and military contests. In the competitive state system of Europe, where there were constant large-scale and decisive military contests, the mechanism was working effectively. In contrast, in an imperial order such as that which existed in China, the mechanism was weak or nonexistent most of the time. Consequently, in noncompetitive imperial order, there were suppression of commerce and the merchant class; state monopoly; excessive regulations; foreclosure of internal and external trading activities; personal rule (versus the rule of law), which in turn leads to the insecurity of property rights; few public services in return for the high level of extraction; and, probably most important of all, state-enforced orthodoxy in thinking, which stifles intellectual creativity and scientific inquiry—these are the reasons that cause empires to have economic failures.[13]

Political military competition generates competitive pressure for states to constantly increase productivity by institutional and technological creativity. The result is a permanently higher economic growth rate. Like Lewis Caroll's red queen in *Through the Looking-Glass*, states in an intensively competitive international political system must put in great effort just to keep up with their diligent and innovative rivals. To get ahead of competitors requires an even higher growth rate through greater innovations and creativity by truly extraordinary gigantic effort. Remember the red queen's answer to Alice? "Now, here, you see, it takes all the running you can do, to keep in the same place. If you want to get somewhere else, you must run at least twice as fast as that!" The competitive pressure to at least keep pace with, if not get ahead of one's rival, in terms of productivity, efficiency, resources and prowess generates a constant stream of gains in higher economic growth and technological progresses and institutional improvements. In the long term, human diligence and creativity are the most important factors in determining progress and prosperity.

---

12 Refer to Kennedy (1981).
13 Wesson (1967) studies the economic failures of imperial orders, especially in chapter 4, 6 and 7. Wesson (1978, pp. 87-90) analyzes the positive effects of interstate competitions on development.

Another important way that military technology affects economic performance is through affecting the resource extraction capacity of state domestically. States with great resource extraction capacity could better finance provisions of public goods, including public intermediate inputs. Furthermore, political military competition provides states the incentive to improve public goods provisions for better economic performance. Tilly (1975, 1992), who shares the view that it was political military competition that drove the European states to modernize, argues that there was a relationship between the scale of international conflicts, the choice of state fiscal apparatus and level of services rendered, and military capacity. The initial spark to the chain of events in Europe was the series of innovations in military technology that increased the economies of scale in warfare from the sixteenth century: the so-called military revolution.[14] Tilly (1975, 1992) attributes the rise of modern European national states to the more frequent and large-scale interstate wars in Europe following this military revolution. States built up bureaucracies and replaced indirect rule with direct rule. The larger scale of international conflicts pressured the states to increase their fiscal, economic and military capability; the rise of largescale standing armies (and navies) and largescale warfare brought forth the bulky modern national states and modern economies. Tilly (1975), put it this way: "War made the state, and the state made war." (42)

Tilly (1975) summarized the European experience in state making and war making in this way:

"The formation of standing armies provided the largest single incentive to extraction and the largest single means of state coercion over the long run of European statemaking. Recurrently we find a chain of causation running from (1) change or expansion in land armies to (2) new efforts to extract resources from the subject population to (3) the development of new bureaucracy and administrative innovations to (4) resistance from the subject population to (5) renewed coercion to (6) durable increases in the bulk and extractiveness of the state." (73)[15]

Tilly (1992) observes that the number of political units in Europe declined from over a thousand in the eleventh century to a couple of dozens in the nineteenth century. The risk of being eliminated from the political landscape was

---

14 Refer to Tilly (1975, 1992) and Parker (1976, 1996), Duffy (1980), McNeill (1982), Dudley (1991), Keegan (1993).
15 Refer to Hintze (1975), Tilly (1975, 1992), Duffy (1980), Cohen, Brown and Organski (1981), Blum and Dudley (1989), Rasler and Thompson (1989), Downing (1992) and Porter (1994).

therefore very high. Tilly (1992) finds that strength of the economy was very important in deciding the outcome of great power rivalry: international political and military competition weeded out inefficient states with weak economies and non-functioning institutions. States were thus pressured to reform inefficient institutions and policies, and to improve both their economic performance and extracting and mobilizing ability.[16]

This book argues that military technology and geography have a great impact on the international political structure, especially the distribution of military and economic capability within the system. The distribution of capability and the economies of scale in warfare jointly decide the level of intensity of political military competition within an international system. If there is a high level of competition, then there will be superior very long term economic performance. A good example is the competitive state system of Europe during the European rise to global economic dominance. On the other hand, if competition is minimal, then there will be economic stagnation or decline. Good examples are ancient Egypt (during the Old Kingdom) and imperial China. Furthermore, the distribution of power within an international political affects not just the intensity of political and military competition but also the risk attitude of a state. When the constituent states of an international system have very unequal power positions, the stronger contestant becomes complacent and risk-averse since he has much to lose and very little to gain by taking risk while the weaker contestant is insecure and risk-seeking as he has much to gain and very little to lose. Extremely high risk-averse and risk-seeking attitudes are not beneficial for long-term development. Contestants with extremely high risk-averse or risk-seeking attitudes will undertake pricey measures to either guard against uncertainty, or to gamble in excessively risky ventures at the expense of average long-term gains of the economy.

Some of the case studies in Bernholz and Vaubel (2004) do not fit well in the original Hume-Kant hypothesis. Specifically, the Hume-Kant hypothesis has difficulty explaining the economic performance of Japan under the Tokugawa Shogunate, the splendid cultural and economic achievements of India during the reign of the Gupta Empire, and the cultural creativity and economic vitality of the Islamic civilization under the Umayyad and Abbasid Caliphates. Bernholz and Vaubel (2004) are also unable to account for the failure of Southeast Asia to be at the forefront of human civilizations despite several potentially-advantageous factors: the continuous functioning of a Southeast Asian state system since the late classical era, the auspicious position of Southeast Asia as a crossroads of major civilizations, and the abundance of resources in Southeast

---

16 Kennedy (1987) has the same finding.

Asia.[17] By studying how military technology affects geopolitics and very long term economic performance, this book is able to account for the above cases that eluded the Hume-Kant hypothesis.[18]

---

17 Refer to Cook (2004).
18 Refer to Cowen (1990) for a related treatment.

CHAPTER 2

Military Technology, Geopolitics and Economic Development

1. Mass Factor and Relativist Concern

The economies of scale in conflict (or the returns to scale in conflict, or military decisiveness, or the mass factor—terms which are used interchangeably in this book), are well-studied by scholars such as Dudley (1990, 1991, 1992) and Hirshleifer (1995).[19] It also occupies a prominent position in many grand strategy theories such as that of Wittman (1991). The mass factor measures the relative advantage a larger contestant has over its smaller rival, where a greater mass factor represents a greater relative advantage. If a larger force has a great probability of victory over a smaller force and could do so at little cost, then there is a large mass factor. On the other hand, if a larger force has about the same probability of victory as its smaller foe and victory could only be secured at the expense of great costs, then there is a small mass factor.

Dudley (1990, 1991, 1992) expounds another way of looking at the concept of economies of scale in conflict: when there are greater economies of scale in conflict, a larger force will suffer fewer casualties when winning victory against a smaller enemy. At the extreme, when the economies of scale in the application of force are infinitely large, a larger force will suffer no casualties when crushing a smaller enemy.

The mass factor is an aggregate technological parameter and does not refer to any specific technological characteristics of the weaponry or the auxiliary system. Examples of increases in the mass factor are the emergence of large standing armies and navies in the 16th and 17th centuries (an emergence that arose due to the military use of gunpowder weaponry which made wars more decisive), and the emergence of the citizen mass army since the French Revolution (as a consequence of rising nationalism which made possible the deployment of troops beyond the personal observational range of officers as nationalistic soldiers are less likely to desert or defect). Technological changes in transportation and infrastructure support systems that enabled a larger body of troops to engage in battles and campaigns increase the economies of scale in conflicts. Likewise, technological advances in communication that

---

19 Refer to Bush (1974) for related treatment.

allowed the coordination of military operations across expansive geographical areas increase the economies of scale of conflict in a geographical sense.[20] Prior to these advancements in national ideology (for example, nationalism), technology and infrastructure, conflict was conducted on a much smaller scale. Furthermore, the very nature of conflict within the medieval era presented the difficulties of taking of castles and fortified areas and caused medieval conflicts to be long drawn out and indecisive siege warfare. Only until the gunpowder military revolution did the nature of warfare change radically and with it the economies of scale in warfare.

In addition, non-military factors affect the economies of scale in warfare. They could make conquest by the stronger of the weaker more difficult. These factors include racial distribution, linguistic and religious divisions, and geographical features.[21] For example, India has many languages and ethnic groups, as well as mountain ranges that run in an east-west direction and separate India into the Northern Indo-Gangetic Plain and the Southern India Peninsula. Consequently, India had a smaller mass factor than China under the same military technology and was more difficult to unite or hold together than China, which had greater uniformity in written language and religion.

The economies of scale in conflict further decide how feasible it is to conquer other states and control conquered territory. If the economies of scale in warfare are greater, then conquests and aggrandizements are more likely to happen and continue to expand to spur on further conquests. Therefore within an international political system there will be fewer constituent units and they tend to be larger in size. On the other hand, when there are smaller economies or diseconomies of scale in conflict, a larger force will suffer considerable casualties when winning victory against a smaller enemy. In the extreme case when the economies of scale in the application of force tend to zero, when trying to overcome a smaller enemy, a larger force will have to suffer as many (or even more) casualties as its smaller foe suffers. Under such circumstances, aggrandizement and empire and state building are more costly and are quite impossible. Therefore, the geopolitical landscape tends to be more fragmented when there are smaller economies or diseconomies of scale in warfare.

Another factor affecting the ease of state building is the combined military and economic efficiency of one state relative to that of its rival. This factor

---

20 Dudley (1990, 1991, 1992) gives the concept of economies of scale in conflict an extensive treatment and applies it to explain the changes in the territorial size of the states. Hirshleifer (1989, 1991, 1995) formalizes the concept and analyzes how it shapes the anarchic or hierarchical condition. Refer to Raaflaub and Rosenstein (1999, pp. 364-368) on theories of expansion and contraction of empires.
21 Refer to Hirshleifer (1995).

measures the relative efficiency of the contestants in conquering and securing contested resources and turning these resources into military capability for further contests. A more asymmetric relative combined military and economic efficiency aids state and empire building and leads to a more uneven distribution of military capability and economic resources. For relative combined military and economic-fiscal efficiency between rivals, efficiency in one area could be enhanced or offset by efficiency or inefficiency in another area. A good example is Sung China which enjoyed overwhelming relative efficiency in the economic area compared to its semi-nomadic semi-agrarian rivals controlling northern China. That advantage however was offset by the overwhelming military efficiency of its semi-nomadic, semi-agricultural northern rivals.

Among the many determinants of relative combined military and economic efficiency is geography. A state controlling an agricultural core area, for instance, will have a greater efficiency in taxation than a state controlling a fragmented hinterland. If there are great economies of scale in warfare, then after rounds of contests, chances are that the state controlling the core area will expand at the expense of the state controlling the fragmented hinterland. This was the experience of war-making and state-making in Europe. A good example was the expansion of the Kingdom of France against the Dukedom of Burgundy.[22]

Another factor is lines of trade, transportation and communication. States controlling important lines of trade, transportation and communication will have greater relative combined military and economic efficiency.[23] For example, the Nile River provided substantial economic benefits and served as an important line of transportation and communication in peace or war for ancient Egypt. With the greater relative combined military and economic efficiency accorded by the Nile River, the ancient Egyptian Empire was unchallenged for a long time. It was only with the inventions and uses of horse drawn war chariots and iron weapons in warfare did ancient Egypt face a significant external threat. The Byzantine Empire is another excellent example. Its control of the Black Sea—Aegean Sea trade route through the strategic positioning of its capital, Constantinople, ensured Byzantine dominance over the Eastern Mediterranean regions for more than a millennium. The Byzantine Empire outlasted its Western imperial counterpart of Rome by close to a thousand years.

Asymmetry in relative combined military and economic efficiency could be military in origin as well. Nomads, for instance, have much greater war efficiency than settled societies when cavalry is an important branch of the military. Many of the well-known gigantic empires were established by nomads through

---

22 Refer to Bean (1973) and Tilly (1992).
23 Refer to Friedman (1977).

conquests of settled societies: good examples are the Umayyad Caliphate, the Mongolian Empire and the Timurid Empire. The heavy infantry military revolution of the classical era, on the other hand, gave settled agrarian societies with abundant manpower great military advantage. Consequently, the Roman Republic with its abundant and politically-committed manpower had far greater relative combined military and economic efficiency than the other Mediterranean powers, including the Carthaginian maritime trading empire, the Greek republics and the Hellenistic states. A similar story happened in China around the same period, with the establishment of the empire of the Qin and Han Dynasties.

In sum, economies of scale in warfare and asymmetry in relative combined military and economic efficiency help concentrate resources and military capability in the hand of the advantaged contestant. On the other hand, decreasing returns to scale in warfare and symmetry in relative combined military and economic efficiency hinder the consolidation of resources and capability in the hand of a single state. Resources and capability are thereby dispersed among the contestants and a pluralistic international order (which might be a state system) is maintained.[24]

The distribution of relative capability refers to how military capacity is distributed among the contestants. The distribution of relative capability and the mass factor jointly determines the intensity of international military-political competition. The intensity of political military competition is captured by the concept of marginal effect of relative capability, which measures the effect of an additional unit of relative capability on the probability of victory in the contest. For convenience, it is also called relativist concern, or competitive spirit, or a concern for relative capability in this book. It is derived from the contest for power. A key feature of power is that it is relative. Since power is relative, the state therefore views military and economic capability as relative, at least to some extent. States therefore care about their relative strength in military, economic, technological or other fields of human endeavor so long as these have implications for their relative power in international arenas.[25]

Two characteristics of the marginal effect of relative capability are of special interest. One is that it peaks when the two rivals have equal military capability. That is to say, the effect of an additional unit of military capability of contestant 1 on the probability of victory of contestant 1 over contestant 2 is at its greatest when the military capability of contestant 2 is equal to the military capability of contestant 1. If the difference in capability is too great, then there

---

24 Refer to Hirshleifer (1995) and Nti (1999, p. 424). Refer to Gilpin (1981) and Hirshleifer (1988, 2000, 2001) for related discussions.
25 Refer to Grieco (1988a, 1988b, 1990), Gowa (1989), Baldwin (1993), Gowa and Mansfield (1993), Grieco, Powell and Snidal (1993) and Gowa (1994) for related discussions.

will not be any real contest and anarchy itself might give way to hierarchy.[26] The more closely matched the two rivals are, the larger the concern for relative military strength. When rivals are equally matched, the incentive to outdo each other is at its greatest. States in the state system where contestants have largely equal military capability are therefore keenly aware of the strategic importance of the relative capabilities of the constituent units. For instance, the concept of balance of power invariably entered the minds of statesmen in the ancient Greek city-state system, the medieval Italian city-state system and the modern European state system. This characteristic of the marginal effect of relative capability curve is what the Hume-Kant hypothesis is about. Another way to look at this characteristic is that an increase in the asymmetry in relative capability reduces the relativist concern or marginal effect of relative capability. For convenience, such a reduction in marginal effect of relative capability is termed a negative asymmetric effect. Conversely, a reduction in the asymmetry in relative capability increases the marginal effect of relative capability. In this case the asymmetric effect is positive.

The second important characteristic of the marginal effect of relative capability is that it is affected by the mass factor. An increase in mass factor has two effects on the marginal effect of relative capability. The first is the scaling effect. Since size now confers greater advantage, there is greater concern for relative military capability. That is to say, given greater economies of scale in conflict, the effect of an additional unit of military capability of contestant 1 on the probability of victory of contestant 1 over contestant 2 is greater. By the scaling effect, given an increase in mass factor, there is an increase in marginal effect of relative capability for both players. The second effect is the unbalancing effect. With greater economies of scale in conflict, the bigger player becomes more powerful and the weaker player weaker. This greater disparity in power between the contestants dampens the competition between them. The unbalancing effect makes both players less concerned about relative capability and reduces the marginal effect of relative capability. That is to say, greater economies of scale in conflict amplify the disparity in power between the contestants and consequently, the effect of an additional unit of military capability of contestant 1 on the probability of victory of contestant 1 over contestant 2 is smaller.

The size of the unbalancing effect depends on the degree of asymmetry in capability between the two contestants. The greater the asymmetry in capability, the greater the unbalancing effect. The unbalancing effect is zero if the two rivals have equal capability since in this situation, their power remains equal whatever the economies of scale in conflict. Therefore, if there is a rough

---

26 Refer to Hirshleifer (1995).

balance in the relative capability of the contestants, then an increase in economies of scale in conflict increases the marginal effect of relative capability since the positive scaling effect dominates the negative unbalancing effect. Consequently, when the military capabilities of the two states are roughly equal, an increase in economies of scale in conflict increases the concern for relative capability and the intensity of contest. Conversely, if there is great disparity in the relative capability of the contestants, then an increase in economies of scale in conflict decreases the marginal effect of relative capability since the negative unbalancing effect dominates the positive scaling effect. Consequently, when the military capabilities of the two states are very unequal, an increase in economies of scale in conflict decreases the concern for relative capability and the intensity of the contest.

Of great importance to the main argument of this book is that when the mass factor is large, as the distribution of capability becomes more asymmetric, the marginal effect of relative capability reduces significantly. In other words, when there are great economies of scale in conflict, (that is, size confers great advantage), as the contestants become more unequal in their military capability, the effect of an additional unit of military capability of contestant 1 on the probability of victory of contestant 1 over contestant 2 reduces significantly. On the other hand, when the mass factor is small, that is, size confers little advantage, as the distribution of capability becomes more asymmetric, the marginal effect of relative capability reduces relatively little. In other words, when there are small economies of scale in conflict such that war is indecisive, as the contestants become more unequal in power, the effect of an additional unit of military capability of contestant 1 on the probability of victory of contestant 1 over contestant 2 reduces relatively little. That is to say, the difference in the concern for relative capability between a state system (in which contestants have largely the same level of military capability) and an imperial order (in which the leading power has the overwhelming share of military capability) is greater when the mass factor is larger. For a mass factor that is extremely small, there is practically no difference between an imperial order and a state system in their relativist concern, for in this case, relative capability has very little impact on the probability of victory in a military contest.

The relative power position of a contestant in a political military contest affects not just his relativist concern but also his attitude towards risk. A very weak contestant is in a precarious position and has little to lose and much to gain. The relative power position induces in such a very weak contestant a risk-seeking attitude towards economic decisions or any decision that might affect his power position. On the other hand, a very strong contestant is secure and has little to gain and much to lose. The relative power position induces in such a very strong contestant a risk-averse attitude in economic undertakings, or in any undertaking that has power implications. The weaker the highly disadvantaged contestant, the more risk-seeking he is. On the other hand, the stronger the highly advantaged contestant, the more risk-averse that stronger contestant is.

The mass factor affects the risk attitude as a larger mass factor accentuates the disparity in power. A very weak contestant becomes more risk-seeking when there is a larger mass factor and is less risk-seeking when the mass factor is smaller, for given the same disparity in capability, the very weak contestant is in a very perilous state with a larger mass factor, but less so when the mass factor is smaller. Similarly, a very strong contestant is more risk-averse when there is a larger mass factor and is less risk-averse when there is a smaller mass factor, for given the same disparity in capability, the very strong contestant is very secure with a larger mass factor and less so when the mass factor is smaller. Therefore, when there is great asymmetry in relative capability and a large mass factor, the very strong contestant is extremely risk-averse while the very weak contestant is extremely risk-seeking.

Extreme risk-aversion and extreme risk-seeking attitude cause great distortions in economic decisions (especially investment decisions), and create inefficiency in allocation of resources. Instead of choosing to maximize expected returns, the contestant engages in either too much risk taking (if he is risk-seeking) or too little risk taking (if he is risk-averse). A society characterized by extreme risk aversion is one in which most risky investments are shunned, lucrative though they might be. There are hardly any innovations since innovations invariably involve risky investment. The economy is characterized by stagnation, constancy of income and lack of creativity.

An extreme risk-seeking attitude is harmful to the long term prospect of the economy too. A society characterized by an extreme risk-seeking attitude engages in all kinds of risky ventures, unproductive or non-lucrative though these ventures may be. Resources are set aside for such gambles instead of more productive pursuits. Innovations and creativity that take place in such a society tend not to be of the productive kind. Wide fluctuations of fortunes characterize the economy, though there are not many real gains in productive capacity. Such a society has wild vagaries in short-term conditions though no real substantial long-term progress.

The effect of relative power position on risk attitude explains why an imperial order that is very powerful and secure is very conservative and lacks innovation and creativity. The all-encompassing empire, given its preponderant relative capability, has everything to lose and nothing to gain in terms of power. Such an empire is therefore very risk-averse for any innovation would most probably undermine the power position of the empire and is quite unlikely to improve it. The larger the mass factor, the more powerful and risk-averse an all-encompassing empire is. Consequently, a major civilization composed of only a gigantic, uncontestable and universal or almost universal empire will exhibit a very different risk attitude when compared to another major civilization that is composed of many equal and independent sovereign states. The civilization with a universal empire will be plagued by risk aversion while the state system civilization will not has such a problem, at least not in its severe form. The effect of relative power position on risk attitude also explains the risky "gambling for resurrection" strategy of states facing greatly adverse conditions in war.[27] The larger the mass factor, the more risk-seeking these strategically disadvantaged states with a precarious chance of survival will be.

2. Geopolitics and Economic Performance

Scholars have long recognized the developmental impact of international political structure. For instance, commenting on the importance of sovereign nation states and the competitive state system to the rise of the West, Wesson (1978) notes:

> "Through history, the most important vehicle of competition has been the sovereign state, the supreme organization of society, the great and enduring culture-creating group with which large numbers can identify. Discovery, innovation, productivity, social discipline, and political order have flourished when the sovereign units (which are small enough to arouse feelings of participation yet adequately large to permit the application of available techniques) have been in competition sufficiently strong to engage emotions yet not so desperate as to destroy the rivals. Yet progress is inherently self-limiting because it does away with the conditions that make it possible. The state systems, by virtue of their inventiveness, have made themselves obsolete. Only that of the West managed to hold out for a millennium by expansion and metamorphosis.' (264)

This section analyzes how military technology and international political structure affect economic performance. A competitive state system is an international political system with a large mass factor and a largely symmetrical distribution of capability and resources. The city-state systems of the classical era had many international political systems with a large mass factor but an essentially symmetrical distribution of relative capability and resources. These systems are good examples of a competitive state system, for instance,

---

27 Refer to Goemans (2000).

the city-state system of classical Greece, China's Spring and Autumn Era, and the pre-Mauryan state system of India. The cause of such symmetry in relative combined military and economic efficiency could be that there are many core areas of roughly equal sizes, resources and productivity.

In all these international political systems there was constant jockeying for power among the contestants. They are aptly described by the term competitive state systems or competitive city-state systems. Armed conflicts severely affect the fortunes of the states. It is not known for sure who will be the victor. Weak states are constantly being eliminated or reduced to vassal status. For instance, eleventh century Europe had over a thousand principalities. By the time of World War One, only a dozen of them remained. A good example of how a state could disappear was the partitions of Poland by Austria, Prussia and Russia. A once major power of Europe was eliminated due to its inability to match the other states in terms of military capability.[28]

Given the greater economies of scale in conflicts, each state is too small to fully exploit the scale economies as they are trying to expand. Control over resources and large-scale organization are important for the capability of the contestants. The expanding states are crowded together and rivalry among them is very intense. There is a tendency for the constituent units to reorganize through wars and other means to become larger, in order to better exploit the economies of scale in warfare. The number of constituent units is therefore decreasing. If the process of competition continues indefinitely or, if the military capability or the relative combined military and economic efficiency of one contestant is significantly augmented relative to the rest, then the competitive state system will end up being an empire. If a constituent state of the competitive state system gains disproportionate capability relative to the rest, then the momentum of empire building will be set in motion. The delicate balance of power of the competitive state system is hard to maintain and easily disrupted.

In a competitive state system, the constituent member states have acute awareness about power and relative capability in military and economic spheres. There is constant and immense pressure for the states to be powerful. The drive for power and survival makes the states strive to outdo each other in every aspect of human endeavor that affects the power of states in the international arena, with wars and military contests serving as the ultimate test of state power. Comparisons with other states help to goad states from complacency and decay. The possibility of defeat in the international arena haunts the governing elite as defeat brings not only humiliation and losses, but

---

28 Refer to Tilly (1992).

the increased likelihood of revolution, coup d'etat and other forms of unrest. Fear of the worst-case scenario—conquest—is the driving force for states to be powerful. There is therefore a strong concern for relative capability.

In a competitive state system, the strong concern for relative capability is a potent developmental force. States try to be more powerful militarily and economically than other states, employing many different measures to enhance their prowess. Entrepreneurial statesmen implement institutional changes to boost the economy and enhance state capacity. Good examples abound in the competitive state system of Europe. For instance, Peter the Great westernized Russia in order to make it a great power. Frederick the Great made Prussia into a centralized military state so as to be able to compete with other European powers. At an earlier time on the other side of Eurasia, the Era of Warring States of China (475 – 221 BC) provides many good examples as well. At times, development within states can be extraordinarily dramatic: entrepreneurial statesmen may seize power through social upheavals and overhaul the institutional framework, changes that may then generate greater state capacity to cope with the pursuit of power in the international arena. The French Revolution, the Russian Bolshevik Revolution and the Japanese Meiji Restoration are good examples.

The constituent states of a competitive state system do not have extremely risk-averse or risk seeking attitude to significantly distort their economic decisions, thanks to the largely even distribution of power within the competitive state system. The strong relativist concern of the state is therefore translated into great developmental efforts without being significantly distorted by an extreme risk-averse or risk-seeking attitude. Consequently, the intense competition prompts the state to intervene extensively and rationally in the economy to secure a strong economic foundation to support the military machine. The extension of justice by the central government, the substitution of indirect rule with direct rule, and the suppression of feudal wars are examples of measures taken. The state provides public intermediate inputs to boost economic productivity. The share of public intermediate inputs in the economy is high. There are therefore great rational development efforts and achievements in the competitive state system.

An uncompetitive state system is an international political system with a small mass factor and a largely symmetrical distribution of capability and resources. Dense tropical rainforests inhabited by primitive tribes and clans provide a good example of an uncompetitive state system. Given the difficult terrain, there are very small economies of scale in warfare. The distribution of resources and capability among tribes and clans is extremely even, as military conquests and political aggrandizements are constricted by natural geography and climate. The constituent units are separated by severe natural barriers which limit conquests and formation of a unified state. Each constituent unit has expanded to its natural boundary and faces a low level of external threat from the others. Geography, rather than capability and organization efficiency, determines the boundaries of the constituent units. The constituent units might not even be contiguous. The relative distribution of military capability among constituent states does not matter much in this environment.[29]

In an uncompetitive state system, there is hardly any struggle among the constituent units for hegemony or supremacy. Nature rather than politics decides the status and power of the states. Armed conflicts are of a small scale and short duration. Weak constituent units or states continue to exist, despite inefficiency in military, economic and other aspects. A good example of an uncompetitive state system was sub-Saharan Africa before the colonial era. There might be some interstate or intertribal conflicts, raids and wars, but given the pre-modern military, transportation and communication technology, as well as the hostile terrain, there was hardly any significant competition among the African states and tribes. Another good example is Oceania before the colonial era. Vast distances across the ocean separated settled societies and states. Though trade and cultural contacts were possible or even frequent, mutual conquests were much more difficult. Consequently, the geopolitical landscape was determined more by geography than by politics.

The uncompetitive state system is rather stable. The conditions for its existence, a small mass factor and an essentially symmetric relative combined military and economic efficiency, are quite common in history. There is no tendency for the system to be united under one empire. Wars might be frequent in the system, but they are likely to be small in scale and, due to the natural geographical conditions, inconclusive—perhaps more raids than wars. State and empire building is difficult given the lesser economies of scale in conflict. The size of the state does not confer much strategic advantage. Control over resources and large organization has no significant effect on the capabilities of the contestants.

In an uncompetitive state system, the constituent member states have little awareness about power and relative capability in military and economic spheres as there is hardly any struggle among them for hegemony or supremacy. Nature rather than politics decides the boundaries between states and the status and power of the states. Weak states continue to exist, despite inefficiency in military, economic and other aspects. Given the highly even distribution of relative capability within an uncompetitive state system, the constituent states do not have extremely risk-averse or risk-seeking attitude that might significantly distort their economic decisions. The relativist concern of the state is therefore being rationally translated into developmental efforts without being significantly

---

29 If there are very low economies of scale in conflict and public administration, then there will be no state. Refer to Oppenheimer (1975), Tilly (1975, 1992), Friedman (1977, 1979), Duffy (1980), Levi (1981, 1988), Best (1982), Blum and Dudley (1989), Blum (1991), Dudley (1991), Wittman (1991), Keegan (1993) and Porter (1994).

distorted by a strong risk-averse or a strong risk-seeking attitude. However, within an uncompetitive state system, there is hardly any pressure for the states to be powerful. There is no strong relativist concern to act as a drive for developmental effort. There is hardly any drive for power among the states and therefore very little concern for relative capability in state preference. States do not try to be more powerful militarily and economically than other states. There are therefore hardly any development efforts and achievements in the uncompetitive state system.

A stable or uncontestable imperial order is an international political system with a large mass factor and an extreme distribution of capability where the strongest contestant almost monopolizes all military capability and resources within the system. The ancient Egyptian Empire during the bronze era and before the invention and use of iron weaponry and horse-drawn war chariots fits the depiction of a stable imperial order perfectly. After the unification of the Nile River valley under the leadership of Upper Egypt, no viable rival existed to challenge Egyptian power for over a millennium.

In a stable imperial order, there are great economies of scale in warfare and extreme asymmetry in relative combined military and economic efficiency. Consequently, in the steady state equilibrium of continuing conflicts, the imperial regime has expanded to the limits of its natural boundary. Given the large mass factor, the control over resources and large organization is important for military capability. Size confers great advantage in warfare. Since the imperial regime has monopolized or almost monopolized both economic resources and military capability, it is difficult for other players to challenge or contest the imperial order, and consequently the imperial order is quite stable.

The stable imperial order is secure and entrenched. If the core empire suffers defeats and a decline in relative capability, the concentration of resources under its disposal and the greater economies of scale in warfare ensure that the core empire will quickly regain its eminent position. If, under extraordinary circumstances, the imperial regime collapses, the interim or transition period between imperial regimes and dynasties is short. Great economies of scale in warfare and great asymmetry in relative combined military and economic efficiency facilitate the swift consolidation of resources and capability in the hands of an early winner. Empire building gains momentum easily and quickly. For instance, the history of the early Egyptian Empire (before the use of iron weapons and horse-drawn war chariots) was essentially monotonous repetitions of dynastic turnovers.

In a stable imperial order, the imperial regime is very well entrenched and secure and it is extremely difficult for small groups of challengers from within and beyond the border to challenge the imperial power, given the large mass factor. Should the empire reigns supreme, an attitude of arrogance and complacency can instill itself. Since the empire is supreme in its geopolitical niche, a strong economy is not needed for supporting the pursuit of power in the international arena. The concern for relative capability in military and economic spheres is almost absent since the empire is all-powerful. The empire therefore feels no need for progress and development. Consequently, the empire extracts resources from the economy for consumption and offers very few public intermediate inputs in returns. The system therefore displays very little developmental efforts or achievements.

In a stable imperial order, given the large mass factor and the extremely high concentration of capability in the hands of the imperial order, the imperial regime has an extremely strong risk-averse attitude while the marginal states and latent challengers have an extraordinarily strong risk-seeking attitude. The extremely strong risk-seeking attitudes and extremely strong risk-averse attitudes severely distort the economic decisions of the imperial regime and the marginal states and are highly detrimental to optimal allocation of resources and development. The system is severely plagued by extreme imperial complacency and conservatism and an extremely low level of rational development effort, given the distortions caused by the extremely strong risk-averse attitude and the extremely weak relativist concern on the part of the imperial regime. The extremely strong risk-seeking attitude and extremely weak relativist concern of the marginal states result in a very low level of rational developmental effort in those states as well.

An unstable imperial order has a small mass factor and an extremely uneven distribution of resources and military capability where the strongest contestant monopolizes almost all military capability and resources within the system. The Srivijaya Empire of pre-modern Southeast Asia fits the description of an unstable imperial order well. Pre-modern Southeast Asia had a very fragmented geography that severely constrained political and military aggrandizement. Dense forests with tropical diseases and vast distances separated by seas made military conquests and unified, centralized political control difficult. Consequently, most of maritime Southeast Asia was in the tribal stage when European colonists first set foot on the islands. Yet, during the late classical and early medieval era, there was a powerful empire in maritime Southeast Asia, the Srivijaya Empire. The Srivijaya Empire was based on southeastern Sumatra, near the coastal city of Palembang, in a strategic location to control the trade between China and India. Given its control over this vital and lucrative maritime trade route, the Srivijaya Empire possessed much higher combined military and economic efficiency than all other maritime Southeast Asian societies. Consequently, the Srivijaya Empire was able to exert its hegemony over present-day Sumatra, Java, Borneo, Malaya, and Riau Islands, and effectively control the Strait of Malacca and the Strait of Karimata. No other nearby state or society came close in terms of capability, resources, or prestige. The language of the Srivijaya Empire, the Malay language, was thus spread throughout maritime Southeast Asia as a lingua franca, a status that it still enjoys today.

In an unstable imperial order, since there are lesser economies of scale in conflict, sheer size does not confer much strategic advantage. Empire building therefore does not take on its own momentum, and the consolidation of resources in the hands of a single power is not easily achieved. The core empire of the unstable or contestable imperial order tends to control less of their known world compared to that of the stable or uncontestable imperial order. The imperial order is more easily contested by rivals from either within or beyond the borders and is therefore quite unstable.

Relative efficiency in military and economic arenas is significant in deciding the fate of the unstable imperial regime. Should the imperial regime become inferior in combined military and economic efficiency, it will be overthrown by the more efficient challenger. If the contestants are equally matched in combined relative military and economic efficiency, then the system evolves into a state system. Should that happen, the system is then an uncompetitive state system, given the small mass factor. The Srivijaya Empire, for instance, quickly faded into mediocrity after changes in maritime trade routes deprived it of its economic supremacy in the region. Another good example was the Carolingian Empire. The heavy cavalry military revolution reduced the economies of scale in warfare. Given the small mass factor, the Carolingian Empire managed to unite Western Europe only briefly under the charismatic leadership of Charlemagne, but quickly dissolved into an uncompetitive state system after his leadership.

In an unstable imperial order, the pressure for the state to boost the economy for greater revenue to support a more powerful military is quite weak. A strong economy is hardly needed for the pursuit of power in the international arena. The state apparatus therefore remains small and non-interventionist. Since the international hierarchy is clearly in place though weakly enforced, the incentive for jockeying of power is not strong. There is therefore weak relativist concern and drive for development. However, despite its control over most capability and resources, the imperial power is still challengeable by other states due to the small mass factor. A smaller state with greater relative combined military and economic efficiency could significantly challenge the imperial regime. If the gap in combined relative military and economic efficiency persists, the small state will ultimately replace the imperial regime. The fate of the imperial regime does, to a certain extent, depend on relative combined military and economic efficiency. Since size does not confer absolute or overwhelming advantage, and relative combined military and economic efficiency is important, there is some pressure for the imperial regime to be militarily powerful and economically efficient. Therefore a certain level of international competition, latent or apparent, still exists. Consequently, the system exhibits a weak but apparently existent level of relativist concern.

In an unstable imperial order, given the small mass factor and the high concentration of capability in the hands of the imperial order, the imperial regime has a very strong risk-averse attitude while the marginal states and latent challengers have very strong risk-seeking attitude. The high risk-seeking attitudes and high risk-averse attitudes severely distort the economic decisions of the imperial regime and the marginal states and are detrimental to optimal allocation of resources and development. The system is plagued by imperial complacency and conservatism and low levels of rational development effort, given the distortions caused by a strongly risk-averse attitude and a weak relativist concern on the part of the imperial regime. The strongly risk-seeking attitude and weak relativist concern of the marginal states also result in a very low level of rational developmental effort.

3. Major Military Technological Revolutions

Changes in military technology have been shaping the geopolitical landscape since the earliest eras. The introduction of bronze weapons around 3000 BC changed the method of warfare in Mesopotamia and Egypt. Massive infantry formations, wielding bronze weapons and protected by bronze armor, in conjunction with archers with composite bows, resulted in a considerable increase in the economies of scale in conflict. In the enclosed space of the Nile Valley, the unification of the Egyptian Empire was achieved rather early, ca 3000 BC, under the leadership of Upper Egypt. In contrast, the open, fragmented terrain and multiple core areas of Mesopotamia maintained its state system for a further six centuries longer than Egypt. Inter-city rivalry intensified and there was a rise in the construction of massive defensive walls. The greater economies of scale in conflict caused territorial expansion of the political units and consequently, a series of empires were established.[30] First was the Akkadian Empire, which reigned from 2334-2193 BC. This was followed by the Empire of the Third Dynasty of Ur (2112-2004 BC) and then the Babylonian Empire (ca 1900-1595 BC). The Old Assyrian Empire ruled from around 1830-1741 BC.[31]

The invention of horse-drawn war chariots provided military advantages to peoples on horseback and led to waves of nomadic invasions upon ancient civilizations from around 1700 BC. The military use of iron further tilted the combined military-economic efficiency towards the nomads and away from settled societies. As iron utensils and weapons were cheap, in comparison to those

---

30 Refer to Dudley (1991, p. 47-76).
31 Refer to Haywood (1997, p. 42-43).

made of bronze, nomads could also afford them. Iron-wielding Hyksos armies defeated copper-armed Egypt around 1600 BC. The more decisive and mobile form of warfare ended the geopolitical isolation of the Egyptian civilization from the rest of the Near Eastern civilizations, thereby creating the Greater Near Eastern international political system.

Around the 9th century BC, horses were bred large enough to allow the emergence of light cavalry on the battlefield. (Light cavalry fights by shooting arrows on horseback while at full gallop, while heavy cavalry depends on the 'shock' of the charge to overwhelm the enemy; light cavalry is only lightly armored or not armored, while heavy cavalry is heavily or fully protected by armor.) Light cavalry units replaced horse-drawn war chariots. This wave of military technological changes resulted in the rise of combined arms legions, where the light infantry made up the bulk of the fighting force and the light cavalry was the important mobile striking force increasing the economies of scale in warfare. The consequence was the emergence of the Pan Near Eastern empires: first it was the Neo-Assyrian Empire, then the Neo-Babylonian Empire and finally the gigantic Achaemenid Persian Empire. In contrast, in the more fragmented or sparsely populated geography of classical Greece, Ganges India and China, warfare was less decisive, and it was the city-state system that emerged first, followed by the territorial state system.

The heavy infantry military revolution of the classical era increased the mass factor. (Heavy infantry relies on engaging the enemy directly in hand-to-hand combat to defeat the enemy, while light infantry fights by skirmishing and delivering missiles to destroy or disrupt enemy formations. Heavy infantry is heavily armored while light infantry is only lightly armored or not armored.) The phalanx formation developed by the Greek city states and then later perfected with the combined Macedonian phalanx and cavalry formations dominated the battlefield. This was then over taken by the introduction of the Roman legions equipped with superior iron weapons and armor. Consequently, the Greek city-state system gave way to the Macedonian Empire, and the Hellenistic state system in turn gave way to the Roman Empire. The rise of the Roman Empire was further aided by the naval revolution related to the use of the triremes. Naval dominance helped Rome to consolidate control over the whole Mediterranean basin. This same process of empire building through use of heavy infantry legions also happened in China. The heavy infantry revolution ushered in the unification of China under the First Emperor, sweeping away the classical state system of China of the Spring and Autumn Era and the Era of the Warring States.[32]

---

32 Refer to Hui (2005).

The heavy cavalry revolution tilted the relative military-economic efficiency between settled societies and nomads back towards the latter. Consequently, from around 300 AD to the gunpowder military revolution, waves of nomads—Turks, Tungusics, Mongols and Arabs amongst them—invaded agrarian civilizations and established states and empires around the Eurasian world.[33] The heavy cavalry military revolution led to the more or less simultaneous collapse or retreat of the classical universal empires in the period 300 to 600 AD. The ascendancy of cavalry relative to infantry reduces the economies of scale in the application of force, because cavalry relies less on numerical superiority to win battles. The Roman Empire was divided into two and the western part collapsed under incessant nomadic assaults.[34] The Gupta Empire of classical India was weakened by nomadic assaults from Central Asia and slowly faded. The Jin Dynasty of China gave up the central plain of North China, which had the dominant share of resources and population, to the nomads and retreated to south of the Yangtze River, where the battle superiority of cavalry was impaired. In place of the massive classical empires came a myriad of tiny states or state-like force-wielding organizations.[35]

Centuries later, the gunpowder military revolution then again raised the economies of scale in warfare. In regions such as China, Japan, India, the Middle East and Central Asia, there was a dominant core area, and the gunpowder empires soon emerged to dominate the political landscape. Medieval fragmentation gave way to the Ottoman Empire in the Middle East and Eastern Mediterranean region, to the Ilkhanate-Timurid-Saffavid Empires and Afshar-Zand-Qajar Dynasties in Persia and Central Asia, to the Delhi Sultanate and Mughal Empire in India, to the Yuan-Ming-Qing Empires in China, and to the Hideyoshi and Tokugawa Shogunates in Japan. On the other hand, in Europe, there were multiple core areas of about the same size. The greater economies of scale in warfare led to the dismantling of feudalism and the rise of national states, but failed to create a pan-European gunpowder empire. During the process of European war-making and state-making, the small and weak states slowly disappeared. For instance, the use of cannon eliminated city-states, such as Siena, that could not afford the expensive, massive and complicated defense fortifications of trace Italian (or star forts) which were better able to withstand siege cannon fire.[36]

---

33 Refer to Grousset (1970).
34 Refer to Dudley (1992).
35 Refer to Dudley (1991, 1990 and 1992) and Keegan (1993).
36 Refer to Tilly (1992) and Parker (1996).

4. Southeast Asia, Africa, America and Oceania

The waves of major military technological changes discussed in the previous section that swept across the Eurasian landmass had however largely left the non-Eurasian cultures untouched or failed to have any significant impact on their geopolitical landscapes before the early modern era. This has to do with the so called Eurasian dominance in human civilizations.[37] Therefore this book will focus its analysis on the major Eurasian cultures. However, before starting the main analysis on the major Eurasian cultures, this section will study the various major non-Eurasian cultures and provide a justification for the Eurasian focus. Specifically, it will seek to understand their performance and the reasons behind their failures to be the pacesetters in the history of human civilization. That is to say, this section will shed light on the phenomenon of Eurasian dominance with the theory of this book.

The first half of this section studies pre-modern Southeast Asia, a major cultural and economic crossroads of the Eurasian world. The second half of this section studies how the lack of political military competition affected the history of the non Eurasian world: North and South America, Africa and Oceania. By bringing in the role of military technology and political military competition, the analysis provides additional insights into the phenomenon of Eurasian dominance over the non Eurasian world.

Southeast Asia is in close proximity to both India and China. Southeast Asia is at the crossroads of East Asia and South Asia. Since ancient times there were cultural influences from China, India and since late medieval times, Japan and the Middle East as well. It was from India and China that Southeast Asia received its writing systems and literature; models and concepts of statecraft and social hierarchy; and religious beliefs. Southeast Asia also served as the middle man and maritime commercial exchange hub for trade between East Asia and South Asia and the Middle East. Southeast Asia was connected by extensive intra-regional trade even before the use of writing began in the region.

Constant cultural contacts and economic exchanges with India and China started very early in the classical era. By 500 BC long-distance trade involved both China and India. Southeast Asian rulers took the initiatives to adopt Indian culture and religion starting around 400 AD.[38] The Maritime Silk Road that linked China with India, the Middle East and Europe went through the seas and lands of Southeast Asia, making Southeast Asia a pivot in the medieval global

---

37 Refer to Diamond (1997).
38 Refer to Heidues (2000, p. 18, last paragraph and pp. 22-23).

maritime trading system.[39] Yet, among the major Eurasian cultures, Southeast Asia was always at the receiving end of cultural influences. Osborne (2000) poses the question this way:

> "Why has the Southeast Asian region, despite its size, played so small a part in the shifts of global power over the past two thousand years?" (2)

Since Southeast Asia was a state system throughout its history, by the Hume-Kant hypothesis, the region should have produced many cultural and developmental achievements to boast of and to contribute to mankind. Yet, throughout its pre-modern history, Southeast Asia had been at the receiving end of cultural and developmental exchanges. There are many explanations for the Southeast Asian failure to be a leader in the history of human civilization: climate, geography, ecology, etc. All might have played a role. However, the most important factor seems to be the lack of intense political-military competition which contributed to the lack of advances in cultural and economic spheres, and thus explains Southeast Asia's failure to be at the forefront of civilization.

Pre-modern Southeast Asia had a very highly compartmentalized geography. Geography rather than international politics determined political boundaries. Jungles, mountains, wild rivers, rough seas and oceans and hot, wet, malarial lowlands formed insurmountable barriers to invaders or settlers. Nearly all of Southeast Asia is tropical, hot and humid and heavily forested, making military operations difficult. During the late classical and early medieval eras, Chinese empires extended only to North Vietnam but ultimately lost that too in the 10th century; the Chola Empire of South India failed to hold on to any foothold for long in Southeast Asia despite its many military actions; Mongol attacks of Vietnam and Java of the second half of the 13th century failed; Ming Dynasty China's effort to re-conquer Vietnam ended disastrously. That is to say, the mass factor of pre-modern Southeast Asia was very small, significantly smaller than that of the major Eurasian cultures given the same military technology.

Before the arrival of the European colonial powers, small mingled tribes and principalities dominated the landscape of these regions. Nature rather than politics defined the geopolitical landscape.[40] Borders were defined by natural barriers such as mountains, forests, swamps, rivers and seas. Many of these borderlands were, and indeed still are, inhabited by minorities outside the state structure. Centralized states were late to develop in Southeast Asia and were not the dominant form of political organization in the pre-modern period.

---

39 Refer to Abu-Lughod (1989).
40 Refer to Fitzgerald (1973, pp. 54-55) and Jones (1981, p. 167).

In most areas, political integration above the village level was rare.[41] Given its difficult terrain, the rounds of military technological shocks that swept through the Eurasian landmass had failed to produce a pan Southeast Asian empire.

There were at times powerful and prosperous regional power centers in Southeast Asia. However, the geopolitical landscape was dominated mostly by myriad minor powers and tribes that occupied and were entrenched in their own geopolitical and ecological niche. Pre-colonial Philippines, for instance, had no state structures, and was sparsely populated with a largely kinship and village based political organization.[42] None of the major centers could expand far beyond their natural niche, much less of establishing a pan Southeast Asian empire. Political competition took the form of border skirmishes and occasional raids into the heartland of rivals, rather than conquests. From the 7th to 13th centuries, there were major power centers in present-day Cambodia, Myanmar, Sumatra and Java. The first Burmese Kingdom known was based in Pagan, and reigned from the 11th century until Pagan was sacked by the Mongols from China at the end of the 13th century.[43] The most famous power centers of this period were the maritime trading empire of Srivijaya (based in Sumatra) and the land-based agrarian Khmer Empire centered in Cambodia.

The first of these power centers, the Srivijaya Empire, was a coastal trading center and a thalassocracy based in Sumatra. It dominated much of Southeast Asian trade from about the 7th to the 13th century. It was not a territorial state nor was it a power that depended much on military might for its rule. It had no clearly delineated territories nor centralized administrative capacity. Its influence was strongest along the Straits of Malacca, the Karimata Strait (between Borneo and Sumatra) and the Sunda Strait (between Java and Sumatra). Beyond that, its influence took the form of tributary and trading relationships. It was weakened by attacks from the Chola Empire of southern India, and by changing trade routes, which deprived it of its most important source of wealth.

The second of these Southeast Asian power centers was the Khmer Empire, centered on the basin of Lake Tonle Sap. It had a highly developed agriculture which supplied the empire with ample manpower and wealth. At its greatest extent, it controlled much of present-day Thailand, Laos, Cambodia and Southern Vietnam. In earlier periods, Funan, Chenla and Champa were smaller power centers in southern Vietnam and Cambodia preceding the rise of Khmer, but whether or not they qualify as "states" is questionable.[44]

---

41 Refer to Heidhues (2000).
42 Refer to Church (2006, p. 125).
43 Refer to Church (2006, p. 110).
44 Refer to Heidhues (2000, pp. 23-34) and Sardersai (2003, pp. 22-50).

After the collapse of the Srivijaya and Khmer empires, the major powers be-tween the 14th and 18th centuries were: Myanmar under the rulers of Pegu and then Ava (near Mandalay) (1364-1752 AD); Vietnam under the Later Le Dynas-ty (1428-1788 AD); the Thai Kingdom of Sukhothai and then the Ayutthaya (or Ayudhya) based on the Chao Phraya River delta (1351-1767 AD); Vietnam, which had gained independence from China in the 10th century and was expanding to the south; the Majapahit Empire, centered on Eastern Java and Bali (1292-ca 1527 AD); and the Sultanate of Malacca (Melaka), centered on the Malay Penin-sula (ca 1400-1511 AD). The Sultanate of Malacca succeeded Srivijaya's role as a maritime trading power, while the other major power centers were land-based and agrarian. The Sailendras and Mataram rulers of Java were the other powers that preceded the Majapahit Empire.[45] Additionally, there was a galaxy of small-er states, some quite powerful. Among the minor powers were the sultanates of Acheh, Brunei, the Malukas and Sulawesi, and the Kingdom of Cambodia.

None of these Southeast Asian powers however could overpower the rest to establish a pan Southeast Asian hegemony or imperial realm, despite changes in military technology throughout pre-modern history. Compared with other ma-jor cultures of the Eurasian landmass, Southeast Asia remained throughout its pre-modern history an international political system dominated by small states and tribes with a low level of political-military competition between the major constituent units, due to the small mass factor. The low-level competition within the system caused Southeast Asia to be a follower of the other major cultures in terms of advances in human civilization, despite its size and resources. All pre-modern major Eurasian civilizations had larger mass factors than pre-mod-ern Southeast Asia. The geography of Southeast Asia was much more fragment-ed compared with that of the major Eurasian cultures. Pre-modern Southeast Asia therefore had lower relativist concern than the major Eurasian cultures, despite its highly equal distribution of capability. Consequently, pre-modern Southeast Asia failed to play a more important role in the history of civilizations.

Like Southeast Asia, the non-Eurasian cultures also failed to be at the fore-front of human civilization. Diamond (1997) argues that the dominance of Eurasia is based mainly on its ecology, biology and natural geography.[46] Eur-asia is laid primarily over the west-east axis, while the North America-South America landmass and Africa have north-south as the main axis. This, together with Eurasia's large area, results in wider continuous ecological areas in Eurasia compared with other landmasses. There were therefore more Eurasian plant

---

45 Refer to SarDesai (2003, pp. 51-62).
46 Crosby (1972, 1986, 1994) has applied the same approach and analyzed the impact of human activities on the global ecology on a world historical perspective.

and animal species suitable for domestication and more opportunities for the peoples of Eurasia to exchange both innovations and diseases.

The east-west orientation of Eurasia allowed breeds domesticated in one part of the continent to be used elsewhere through similarities in climate and the cycle of seasons. Australia, in contrast, though it also has an east-west orientation, suffered from a lack of useful animals due to mass extinctions. Due to the north-south main axis of the North and South American landmass, the peoples of the Americas had difficulties adapting crops domesticated for use at other latitudes. Africa was fragmented into many economic sub-areas due to the extreme climatic variations from north to south: in both Africa and the Americas, plants and animals that flourished in one area never reached other areas where they could have flourished, because they could not survive the intervening environment. Therefore domesticated plants and animals and technology spread much faster in history inside Eurasia compared to other continents. Hence Eurasia was able to support larger, denser populations, which made trade easier and technological progress faster than in other regions. These economic and technological advantages eventually enabled Eurasians, and ultimately Europeans, to conquer the peoples of the other continents in recent centuries.

The presence of large animals capable of being domesticated to be raised for meat, work, and long-distance communication and transportation further enhanced the advantage of Eurasia relative to other landmasses. The five most useful species of domesticated large animals- cow, horse, sheep, goat, and pig are all indigenous to Eurasia. This livestock helped with the spread of agriculture and eventually cities, and consequently, Eurasia had denser populations with a higher level of trade. There were also more people living in close proximity to livestock. Diseases were transmitted more easily and so natural selection forced Eurasians to develop immunity to a wide range of pathogens. Therefore, when Europeans made contact with the Americas, European diseases ravaged the Native American population, rather than the other way round. So it was easier for relatively small numbers of Europeans to conquer much larger indigenous populations.

Two of Diamond's (1997) arguments are especially relevant to the main argument of this book: the main axial orientations of the continents and the availability of large domesticated animals for long-distance transportation. The combined continent of the Americas is the largest of all the non-Eurasian regions, about four-fifth the size of the Eurasia land mass. Supposedly the combined continent of the Americas should be a close rival of Eurasia in the progress of human civilizations. Yet, the reality is far from that. The main reason is that the main geographical orientation of the Americas is from north to south. This made travel and conquest difficult during pre-modern times as the travelers or conquerors had to overcome differences in climate. This, together with the lack of large domesticated animals for transportation, resulted in lower economies of scale in warfare. There was no contact between the civilization in

Peru and those in Meso-America.[47] Therefore, once the Aztec and Inca empires were established in Meso-America and the Peruvian highland, there was hardly any political-military competition in America. The lack of any significant political-military competition resulted in a very low relativist concern. Furthermore, metallic weapons were not used in warfare in the Americas until the arrival of the Europeans. Warfare therefore had very low military decisiveness and relativist concern was very low consequently. The very high concentration of capability in these two empires resulted in highly risk-averse power-induced risk attitude too. Consequently, complacency and conservatism plagued the pre-modern American civilizations and there was no drive for further progress.

Africa too has a north to south main geographical orientation. Pre-modern sub-Saharan Africa was covered by thick forests, and there were diseases conquered only recently by modern medicine. Some of these diseases were not even conquered yet. All these obstacles resulted in a small mass factor. Therefore civilization and state building started very late, despite Africa's proximity with the first civilizations of the ancient Near East. The pre-modern states or empires in Africa were localized and not contiguous with each other; the political units interacted sporadically and marginally. Due to the very small mass factor, pre-modern Africa had a very low relativist concern. Consequently, there was very little drive to propel the progress of civilization in Africa.

Australia and Oceania were characterized by external and internal isolation in pre-modern eras. In fact, these regions were the most isolated of all major regions. A sparse population was separated by vast distances, which resulted in very low economies of scale in warfare. Political-military competition was almost nonexistent. Most of the region was still at the hunter-gatherer stage when incorporated into the Europe-centered modern world state system in around the 18th century. The extremely low mass factor generated extraordinarily low relativist concern, even though the distribution of resources and capability was highly even. Military technology was still at the pre metallic stage too which further contributed to an extremely low relativist concern. Consequently, there was practically no drive generated by political-military competition to propel the progress of civilization.

In sum, the brief survey of the experiences of Southeast Asia, Africa, the Americas and Oceania reveals the importance of political military competition in affecting advances of economy and civilization. The lack of political and military competition explains the failures of these diverse, large and resource-abundant regions to set the pace of civilization in human history. The result is the so called Eurasian dominance in human civilizations.

---

[47] Refer to Diamond (1997).

5. The Three Strands of Hume-Kant Hypothesis

Political military competition pressures states to keep pace with if not over-take one another in economic performance and technological and institution-al improvements. The other two mechanisms of Hume-Kant hypothesis also bring forth such beneficial outcomes. Cultural diversity and exchanges, eco-nomic freedom and mobility of factors of production across national bound-aries and the liberal constraints that such freedom and mobility imposed on the predatory tendency of the state, enhances economic performance and technological and institutional progresses as well. Though focusing on the ef-fects of military technological changes and political military competition on geopolitics and economics, this book is not espousing a mono-causal theory of political military competition determining economic performance. Other factors and mechanisms definitely have their effects too. Rather, the emphasis on political military competition is for expositional convenience and clarity. Readers interested in the working of the other two mechanisms might want to refer to Bernholz et al. (1998) and Bernholz and Vaubel (2004) and the works cited there.

Another reason for focusing on political military competition is that the work-ing of the other two mechanisms of Hume-Kant hypothesis, namely, cultural and institutional creativity and diversity and, factor mobility across international boundaries, also critically depends on military technology and political military competition between states. There are two major preconditions for cultural di-versity and exchanges, and institutional and policy innovations and imitations to work. Firstly, there must be more than a single state around. Secondly, there must be significant competition between states for them to be interested to innovate to get ahead of other states, or to imitate to catch up or stay on par with other states. Military technology, by affecting the geopolitical landscape and the intensity of political military competition between states, significantly decides the decree of cultural innovation and competition between states.

Military technology, by affecting the geopolitical landscape, influences the number and size of states within a particular geographical region. That fun-damentally decides the possibility of mobility of factors of production across state boundary. There will be greater mobility of factors of production across state boundaries if there is keen policy competition between states to lure more factors to their territory so as to enlarge the tax base, which would oc-cur if there is intense political military competition.[48] Political military drives states to trade and increase their resources and capability. A good example

---

[48] Refer to Alesina, Spolaore and Wacziarg (2000).

was that mercantilism became the guiding principle of many states in the early modern competitive European state system. Another good example was China during the Spring and Autumn Era and the Era of the Warring States when China produced its own mercantilists, including Guan Zi and Lu Bu Wei. Under the premiership of Guan Zi, the Kingdom of Qi during the Spring and Autumn Era actively encouraged commerce and gave merchants many preferential treatments. These policies laid down the economic foundation for the Kingdom of Qi to emerge as the first of a series of hegemonic powers during the Spring and Autumn Era. Similarly, the policies of premier Lu Bu Wei during the Warring States Era greatly enhanced the power of the Kingdom of Qin and facilitated the unification of China under the First Emperor. Another good example of how the concern for power affected the trade policy of the state was Sassanian Persia's conquest of Yemen. The move had the intended purposes of blocking the Roman-Indian trade, to weaken the Roman Empire and profit the Sassanian Empire.

A further reason for focusing on political military competition is that there are reasons to believe that it has a stronger effect on the very long term economic performance of a major culture than the other two mechanisms of Hume-Kant hypothesis. Transportation and communication means and technology are necessary for the working of all three mechanisms, and factors that contribute to one often will help in the working of the other two too. In terms of requirements of transportation and communication capacity, cultural and economic exchanges are less demanding than political-military conquests. Military conquests require not only movements of troops and supplies across geographical space and barriers, but also overcoming the resistance of the defending forces. Therefore, the existence of severe political-military competition means that if states are already in such close proximity, given the technological and geographical conditions, cultural and economic exchanges will be at a very high level, so long as state policies and political factors do not inhibit them. All major cultures therefore have had substantial economic exchanges within their respective geographical regions throughout known history. It was the type of international political structure that differed across time. If a high level of political-military competition existed, then so would a high level of cultural and economic exchange. Therefore a priori theoretic reasoning points to political-military competition as the main force driving the variations in very long term economic performance through time.

The case of pre-modern Southeast Asia and the other cases of non Eurasian cultures provide the empirical justification for the emphasis on the political-military competition strand of the Hume-Kant hypothesis. Since pre-modern Southeast Asia and many of the other non-Eurasian cultures (such as those in Africa and America) enjoyed a high factor mobility amongst their constituents units and yet failed to be at the forefront of human civilization historically, there is therefore a good cause to focus on the political military competition mechanism to explain variations in very long term economic performance among the major cultures, especially those of the Eurasian rim lands. To this task the remaining chapters of this book devote themselves to.

CHAPTER 3

Ancient Middle Eastern Leadership

1. Introduction

Military technology shaped geopolitical structure and economic performance since the dawn of civilization. The pre classical era ancient world provides five case studies for understanding how military technological changes lead to variations in international political structure and economic development. These are the differing city-state systems of Mesopotamia and Egypt after the invention of writing, but before the use of bronze weapons, then the Mesopotamia state system and the Egyptian imperial order after the introduction of bronze weapons and finally, the greater Near Eastern state system after the introduction of horse-drawn chariots in warfare.

The first civilizations of Mesopotamia and Egypt and the Greater Near East are the earliest international political systems in world history. While the early Mesopotamian and the later Greater Near Eastern systems bear resemblance to the early modern European Westphalian system with their equal distribution of capability among the major contestants and emphasis on balance of power, the Egyptian system provides a sharp contrast with its extremely high concentration of capability within the hands of the one imperial power which is often referred to as "the paragon of imperial order".

2. Writing

Developmental puzzles are not unique to the modern world as many also exist from the ancient and classical world. Long before the rise of industrial Europe, there were the first civilizations of Mesopotamia during the fourth millennium and Egypt at the end of the fourth millennium (ca 3500-3000 BC).[49] Next was the puzzling leadership that Mesopotamia had over Egypt after the establishment of civilization. Another was the swiftness of the acceleration of development in several different regions: around 900-700 BC, China, the

---

[49] The Indus-Harappan civilization is not included. It was destroyed early by Aryan invasions. Historians still know quite little about the history of the Indus civilization.

Indian Ganges River valley and Greece all produced surprisingly sudden and splendid classical cultures.

In Mesopotamia, the first cities emerged around 4300-3100 BC. Mesopotamia was the very first civilization in human history, and saw the emergence of the first agriculture and writing systems. The most basic fundamentals of civilization came into existence there: agriculture, writing and urban living. During this formation period, culture progressed rapidly: villages coalesced into cities and cities developed into city-states.

Mesopotamia had an arid landscape, which in a way served as an advantage because construction of irrigation networks stimulated the growth of bureaucracy, cities and states.[50] The emergence in 4000-3500 BC of the Sumerian city-state system, in southern Mesopotamia, was probably the greatest turning point in human history. The earliest known writing originates in the Sumerian city of Uruk (in today's Iraq), around 3400 BC.[51] The invention of writing aided administration, facilitated management of a complex society, and drove the rise of cities.[52]

Parallel to and greatly influenced by Mesopotamia, Egypt also surged forward in the period 3500-2800 BC. Benefiting from trade and technological influence from Mesopotamia, the Egyptian culture crossed the critical threshold of civilization within a very short span of time. From its modest pre-written history period, the Egyptian civilization rose suddenly from the desert into its monumental grandeur. The speed of emergence of higher civilization in the Nile Valley is amazing. This period saw villages quickly develop into cities and cities into city-states—and relatively soon, city-states gave way to empire in the Nile Valley. Chiefdoms, towns and city-states appeared by 3300 BC; cities, state formation and organized technology suddenly came together in the last centuries before 3000 BC. The hieroglyphic script was developed during this era. Egyptian civilization coalesced quickly into an empire with only a very brief city-state phase.

The invention of writing was critical for the rise of civilizations. Writing facilitates the storage of information beyond human memory and allows communication beyond face to face distance. Through writing, communication of information becomes easier; people no longer have to rely upon the cumbersome methods of communicating through face to face exchanges by means of spoken words or gestures. The efficiency gains due to writing allowed the rise of complex societies beyond that of simple agrarian villages. Society could now be organized on a larger scale. The invention of writing increased the organizational capability of human societies, enabling a higher level of specialization

---

50 Refer to Wittfogel (1957) and Dudley (1991, pp. 20-21).
51 Haywood (1997, pp. 16-17, 40-41).
52 Refer to Dudley (1991, pp. 36-43).

of labor and increased productivity. Together with the appearance of writing came the emergence of a professional priestly class who were the first full-time bureaucrats.

In Mesopotamia and Egypt, with the rise of the first cities after the invention of writing, human civilizations spurted forward with great unprecedented speed. In fact, the speed of development during this early dawn of civilization was rarely matched or surpassed in later eras.[53] The efficiency gains due to the new techniques of information storage and communication were an obvious and important factor behind that speed. Another factor was that Mesopotamia and Egypt (after the invention of writing, but before the use of bronze weapons) had a highly equal distribution of capability among the city-states. Interstate conflicts with the pre metallic military technology were indecisive. Military indecisiveness hindered the consolidation of capability and resources in the hands of an imperial power. Consequently, capability and resources were evenly distributed among the many constituent city states of the Mesopotamian and Egyptian systems. Within the narrow confine of the Nile River valley and the uninterrupted terrain of the land lying between the Euphrates and the Tigris Rivers, the highly even distribution of capability among the first city-states generated a fair level of political military competition and relativist concerns despite the low economies of scale in warfare. Furthermore, the lack of high concentration of capability also prevented the emergence of either extremely risk-averse or risk-seeking power-induced risk attitude. Therefore, there was consistent rational developmental effort and the first civilizations were propelled powerfully forward.

In sum, during this formation period of the first civilizations, material conditions improved greatly with sudden speed. Human civilizations in Mesopotamia and Egypt rapidly developed. Of special interest was the extreme speed of development of Egyptian culture during its earliest pre-dynastic era, a speed which was not repeated in the succeeding dynastic eras of Old and Middle Kingdom. The extreme speed of development of Egyptian culture during the pre-dynastic era was partly due to the very compact geography of the Nile Valley. In the narrow confines of the Nile Valley, competition amongst city-states was especially intense. It was this competition that propelled the sudden acceleration of Egyptian civilization during its formation period. This exceptional Egyptian energy was lost when the use of bronze in warfare brought forth the first empires among which the earliest and most perfect was the Egyptian Empire.

---

53 Refer to Wesson (1967, 1978).

3. Bronze

The use of bronze started in Mesopotamia around 3000 BC. It started only very slightly later in Egypt, also around 3000 BC. Bronze weapons are superior to pre metallic weapons (such as stone and wooden weapons) in strength, sharpness and durability. Consequently, battles became more lethal. The use of bronze weapons and armories led to a more decisive and larger scale form of warfare than conflicts with stone or wooden weaponry. Armies grew in size, with infantry wielding sharp-edged bronze weapons and protected by bronze body armor being the mainstay fighting force.

The use of bronze weaponry affected the relative combined military and economic efficiency between contesting city-states in the Nile Valley and, separately, in Mesopotamia. As bronze was expensive and not easily available, states which were richer and commanded more resources and manpower could now field more and larger infantry legions. These legions were standing armies equipped with expensive bronze weaponry. Consequently, these favorably endowed states conducted war more effectively and enjoyed greater relative combined military and economic efficiency in comparison with other less en-dowed states. Bronze weaponry also increased the mass factor. The combined effect of a larger mass factor and a more asymmetric relative efficiency was a higher concentration of capability and resources and the undermining of the city-state systems of Mesopotamia and Egypt. Large-scale territorial states and empires began to emerge in Mesopotamia around 2800 BC while Egypt was united under one empire by around 3100 BC.

The changes in Mesopotamia will be examined first. There are two definitions of Mesopotamia: the first narrower definition refers to the land lying between the Euphrates and Tigris rivers, namely the present-day Iraq. The second broad-er definition refers to the land that lies between the Zagros and Anti-Taurus mountains in the north and the Arabian plateau and Persian Gulf to the south. This area corresponds to eastern Syria, southeastern Turkey, southwestern Iran and the present Iraq. Compared with the Nile River valley, Mesopotamia, in par-ticular greater Mesopotamia, has a more open, complicated and fragmented geography with multiple core areas.

A large quantity of bronze weapons was produced and used in Mesopota-mian the period 2900-2334 BC. The Sumerians pioneered the use of bronze weaponry. The battlefields of Mesopotamia were dominated by armies consist-ing of infantry, wielding bronze weapons and protected by bronze armor, and archers with composite bows. The military technological changes intensified intercity rivalry. Cities erected massive defensive walls to protect themselves. There was a considerable increase in the scale of warfare. Initially, before the emergence of the Akkadian Empire in Mesopotamia in the twenty fourth cen-tury BC, the highly symmetrical distribution of capability of the Mesopotamian city-state system was preserved. The enlarged mass factor due to the invention and use of bronze weapons intensified political military competition without seriously affecting the fragmented geopolitical structure. Consequently, relativist concern increased. Furthermore, given that the even distribution of capability was preserved, relativist concern was elevated without generating either highly risk-averse or risk-seeking power-induced risk attitudes. Development therefore sped up due to the higher relativist concern and without the distortions caused by extreme risk-averse or risk seeking power-induced risk attitudes.

The greater economies of scale in conflict led to territorial expansion of the city-states. Territorial states replaced city-states, and a series of empires was established.[54] Among the earliest conquerors was Sargon the Great of the Akkadian Empire, 2334-2279 BC. This was the earliest empire in ancient Mesopotamia, appearing about six centuries after the establishment of the Egyptian Empire.[55] The Akkadian Empire reigned from 2334-2193 BC. Subsequently, the Empire of the Third Dynasty of Ur ruled between 2112-2004 BC, known also as the Sumerian Renaissance. The Old Babylonian Empire governed ca 1900-1595 BC, and within that span of time saw the rise and fall of the Old Assyrian Empire (1830-1741 BC).[56]

Given the open terrain of Mesopotamia, these empires were more regional hegemonic powers than truly imperial orders. They were unlike the secure imperial order of the Egyptian Empire. They had not truly monopolized, or even come close to monopolizing, military capability within their international political system. The largest of this series of empires, the Akkadian Empire under Sargon the Great, failed to conquer Canaan and failed to subdue either the Hurian kingdoms of the Khabur River valley or the Gutian tribes in the Zagros Mountains. Sargon the Great and his successors constantly faced internal revolts and foreign challengers. Moreover, the Akkadian Empire collapsed within a century, mainly due to the attacks from the Gutians in the central Zagros Mountains.

More often than not Mesopotamia was a field for great powers jockeying for dominance. The more important players were the Elamites of coastal Persia, the Babylonians or Sumerians or Akkadians of Lower Mesopotamia, the Assyrians of Upper Mesopotamia, and the Hittites of Anatolia. The constant great power rivalry of the Mesopotamia system drove the geographical expansion of the system as contestants sought greater power through alliance with external players, overseas colonization, or trade. New components were constantly

---

54 Refer to Dudley (1991, pp. 47-76).
55 Refer to Dudley (1991, p. 66).
56 Haywood (1997, pp. 42-43).

being added: for example, the Mycenaean civilization of Greece became in-volved. Diplomacy and statecraft advanced to a very high level of sophistica-tion, with a good grasp of the concept of balance of power and good use of alliance formations to check potential rivals.

Unlike Egypt, which entered into the imperial era and developmental stag-nation soon after the foundation of its civilization, Mesopotamia retained first its city-state system and subsequently its territorial state system much longer. Consequently, Mesopotamian developmental momentum was preserved for a longer sustained period.[57] Since Mesopotamia after the introduction of bronze weapons produced no all-encompassing empire with an extreme concentra-tion of capability, the greater economies of scale in warfare actually increased the intensity of political military competition and relativist concern within the pluralistic Mesopotamian system as the positive scaling effect dominated the negative unbalancing effect and the negative asymmetric effect. Furthermore, the absence of an all-powerful empire monopolizing capability within the sys-tem preempted the emergence of an extremely risk-averse power-induced risk attitude and its associated economic distortions.

The higher relativist concern and moderate power-induced risk attitude of the Mesopotamian system, when compared to that of the Egyptian imperial order with its low relativist concern and highly risk-averse power-induced risk attitude, manifested itself in many Mesopotamian achievements. Mesopotamia exhibited greater speed of development as well as a high level of creativity in all fronts. The city-state system and the succeeding territorial state system of Mesopotamia produced significant developmental achievements. Mesopota-mia developed the cuneiform script, the first system of writing, which allowed scribes to record business transactions, legal documents, lists of items, and other articles. Mesopotamia also invented libraries and recorded histories. The region also created alphabetical writing, a system of writing taken up by oth-er cultures and from which all alphabets used in the world are derived. Other achievements include the first wheeled vehicles, the potter's wheel, the first surviving codes of law (promulgated by Urukagina, King of Lagash, around 2350 BC), the use of bronze in production, and of course, the first city-states.

On the literary front, Mesopotamia produced much literature of high val-ue. Great religious literature included Enuma Elish and Gilgamesh. The Code of Hammurabi (ca 1792-1750 BC) stands as one of the greatest pieces of early juridical literature. Although Hammurabi's Code is not the first code of laws (the first records date from four centuries earlier), it is the best-preserved legal

---

57 Refer to Wesson (1967, 1978) and Bernholz (1998, pp. 114-121) in Bernholz et al. (1998).

document reflecting the social structure of Babylon at the time. The Code has two hundred and eighty two laws concerning a wide variety of abuses. There were other legal codes that the state further perfected, indicating the state's interest in furthering economic and other development.

On the scientific and technological front, much of mathematical and astronomical science owes its beginnings to the Mesopotamians, who invented the sexagesimal system, used at the time for all types of calculations, and still in use internationally today to tell time. The Pythagorean law was applied as early as the 18th century BC, though it was not yet formulated. The Mesopotamians also developed advanced metallurgy techniques for working with bronze, lead, silver, gold and iron. Although iron did not supplant bronze as the main metal for tools and weapons until around 900 BC, it was in widespread use by 1200 BC, a date generally accepted as the start of the Iron Age.

Mesopotamia had a more advanced level of broadly defined statecraft, as exemplified in the perfection of legal codes, than Egypt did. The Assyrians, for example, throughout the duration of their empire, contributed greatly to mathematics, the military, city planning, governmental administration, architecture and the sciences. Besides possessing military supremacy, Assyrians also had unmatched artistic talent.

The story in Egypt is very different. From the beginning of its civilization to 1595 BC, Egypt was largely isolated from the political and military competition of the Near East.[58] Protected to the north by the Mediterranean Sea and the east by the Red Sea, to the west by the great Sahara Desert and to the south by cataracts along the river Nile, Egypt was a single-river valley civilization cut off almost totally from the rest of the world in political and military affairs: although there were cultural connections with Mesopotamia by trade and general diffusion, politically Egypt stood alone for hundreds of years. The Egyptian civilization was over 1,300 years old before suffering its first major foreign invasion.

The great Nile River defined the Egyptian civilization. The Nile floodplain was the most favorable area for agriculture anywhere in the ancient world. Egypt had little need for irrigation or flood defenses. The Nile also served as Egypt's main highway for both communication and transportation, whether of commercial commodities or troops. In sharp contrast to Mesopotamian culture, which ultimately spread civilization throughout all of greater Mesopotamia and then beyond, Egyptian civilization spent its history focused almost entirely on the Nile River and its desert vicinity.

Within the compact geography of the Nile Valley, the use of the first metal weapons, such as the copper axe instead of wooden clubs, increased

---

58 Refer to Humber (1980, pp. 36-7) and Dudley (1991, p. 11).

tremendously the effectiveness and economies of scale of warfare. Consequently, metal weapons contributed to the uniting of Egypt, around 3100 BC.[59] The isolated and compact geography of the Nile Valley was best suited for the creation and maintenance of an imperial order: the initial city-state system lasted only a very brief time.[60] The imperial regime practically monopolized capability within the Egyptian system and was stable and non-contestable. Egyptian history, before the incorporation of Egypt into the Greater Near Eastern State system, was a monotonous routine of dynastic successions. The extreme concentration of capability in the hands of the imperial regime resulted in an extremely low level of relativist concern as well as an extremely risk-averse power-induced risk attitude. Imperial complacency and conservatism were the hallmarks of Egyptian civilization during the Old and Middle Kingdom eras. Religion controlled every facet of Egyptian society and the priesthood played a very important role in the command economy of Egypt, with very little room for market forces and individual initiatives. Coin money, for example, was introduced into Egypt by foreigners only in the Late Period.

Mesopotamia, in contrast, had multiple core areas and was much more open to external influence. Consequently, civilizations succeeding the Sumerians operated either under a city-state system or a territorial state system, with hegemonic powers of different degrees of dominance, whilst the Egyptians remained firmly under the grip of an imperial regime.[61] The different political-military environment generated different developmental achievements. That difference caused the ancient Middle Eastern leadership in early civilizations.

Mesopotamian leadership in the early phase of human civilization was despite the fact that Egypt had many natural advantages compared to Mesopotamia. The effects of such differences on innovation and development were rather obvious. Egyptian development began to slow down once the First Dynasty united Upper and Lower Egypt around 3100 BC, while the Mesopotamian civilization retained its momentum much longer. This so-called ancient Middle Eastern leadership in the early phase of human civilized history was from ca 3500 to 1000 BC.

The low relativist concern of Egypt was manifested in the cultural heritage best known to the world: pyramids and the other gigantic monumental buildings and Egyptian arts associated with these buildings. In these areas the Egyptian civilization outdid Mesopotamia, though Mesopotamia has more important heritage

---

59 Refer to Dudley (1991, p. 55).
60 Refer to Dudley (1991, pp. 54-55).
61 Hicks (1969, pp. 19-20) refers to ancient Egypt and China as almost perfect imperial orders.

in most other aspects. The amount of resources that the Egyptian Empire poured into such economically unproductive projects showed the lack of concern for relative power and development. Essentially, Egypt could afford to waste its resources on grandeur that would endure through the ages—but in Mesopotamia those resources were used more productively, if less ostentatiously.

As already suggested, a prominent example of the leadership of Mesopotamia over Egypt in early human civilization is the development of laws. According to legal historians such as Morris (1911) and Zane (1927), ancient Egypt is not in the main line of development of laws that the later world inherited. Mesopotamia, in contrast, bestowed upon later civilizations its heritage of laws, including commercial law. The laws developed by Mesopotamia were received and further developed by the Phoenicians, the Greeks and the Romans and came to define the legal framework in use around almost the whole world today. The greater emphasis on economic development by Mesopotamia was very clear.

In sum, Egypt devoted a very high share of its resources to monumental burial buildings and produced no significant law codes. Mesopotamia, on the other hand, produced many law codes of great significance in the development of law. This contrast shows that the Egyptian elite cared more about their afterlife, while Mesopotamian leaders had to worry about rival powers and constantly catered to the needs and interests of important constituencies such as the merchants. The very different attitudes of the leaderships of the two civilizations had a major impact on the development and achievements of their civilizations. Consequently, Mesopotamia produced many great literary works including epics. Yet Egypt, despite its more peaceful and prosperous existence, can boast of no epic literary tradition. Egyptian science and technology were also far less advanced than those of Mesopotamia.

4. Chariots and Iron

Horse-drawn war chariots started to be used around 1700 BC. War chariots gave a significant military advantage to nomadic peoples, as exemplified by the Hittite sacking of Babylon in 1595 BC. The adverse shift in battlefield advantage against the settled agrarian civilizations, and in favor of the nomads, ushered in a dark age of approximately two centuries.

Then iron was introduced into warfare. The iron weapons were much cheaper than those made of bronze and therefore were widely available to the nomads. The nomads thereby gained further military advantage versus settled civilizations. Starting around 1250 BC, there was another period of destruction. The waves of attacks by nomads against the ancient civilizations took place around 1250-1150 BC, including the famous raids by the "Sea Peoples". The Trojan War immortalized by Homer in *The Iliad* was one of these raids.[62]

All ancient civilizations suffered from nomadic invasions and quite a number of them completely collapsed. From around 1700-1400 BC, nomadic invasions destroyed the Cretan and Indus civilizations. Hyksos armed with composite bows and horse-drawn war chariots defeated the copper-armed Egyptian infantry ca 1600 BC, whilst ca 1400 BC, Mycenean raiders invaded Greece. Greece was again invaded around 1200 BC by Dorians, leading to the so-called Greek Dark Age. Nomadic conquests led to Hyksos rule in Egypt and Kassite rule in Mesopotamia.[63] Further to the east, in India, Aryan invasions destroyed the Indus civilization, which was never rebuilt.[64] In China, nomadic chariot warriors conquered the Yellow River valley and established the Shang dynasty (1525-1028 BC).

The nomads built empires with their horse-drawn war chariots and used these vehicles to extend the reach of their empires. Due to the more mobile war chariots and the more battle effective iron weapons, an international political system comprising the whole Middle East emerged. The geopolitically isolated and distinct systems of the first civilizations of Mesopotamia and Egypt were subsumed under this new larger system.

These changes especially had great effects upon the Egyptian civilization. The invention of war chariots led to the 1640 BC Hyksos invasion of Egypt. Bronze weapons, war chariots, composite bows, and scale armor were among the weaponry used. Only in 1531 BC did Egypt manage to expel the Hyksos. For the once-isolated Egyptian Empire, from then on, there were more contacts with external influences. Responding to the new international political and military environment with its more mobile and more decisive kind of warfare, Egypt expanded into Nubia and the Levant for control of more resources and forward defense.

This expansion of the Near Eastern system occurred during the era of the Hittite and Middle Assyrian Empires (1595-1000 BC). The major powers were the New Kingdom of Egypt, the Hurian Kingdom of Mittani (which controlled present Syria and Lebanon), the Hittite Kingdom of Hatti (which controlled central Anatolia (modern Turkey)), the Assyrian state (present central Iraq), the Babylonians (based in present southern Iraq) and the Kingdom of Elam (which controlled present southwestern Iran).

The Greater Near Eastern international political system had a fragmented and open geography with multiple core areas, as the new technology of

---

62 Refer to Sandars (1978, pp. 190-191).
63 Refer to Stavrianos (1982).
64 Refer to McNeill (1999, pp. 32-35).

chariots increased the projection capability of military power across geographical distance and vastly expanded the original Near Eastern system. Egypt, once in a separate system of its own, was then included. The geography of multiple core areas caused a rather symmetrical relative combined military economic efficiency. The multiple core areas and open terrain of the new Greater Near Eastern international political system also resulted in a smaller mass factor than what Egypt had previously faced. The new mass factor was smaller than that which had existed in the compact geography of Egypt given the same or even older military technology. Consequently, the distribution of resources and capability within the Greater Near Eastern international political system was quite even. The Greater Near Eastern system was a multi-polar state system most of the time.[65]

The incorporation of Egypt into the Greater Near Eastern state system resulted in a more equal distribution of capability among the constituent states. Hegemonic powers in Mesopotamia were now balanced by the might and resources of the Egyptian Empire. The more even distribution of military capability and resources among the major powers generated greater relativist concern and developmental drive, especially for Egypt which was freed from its geopolitical isolation.

The induction of Egypt into the Greater Near Eastern system also affected the Egyptian power-induced risk attitude. The more equal distribution of capability of the Greater Near Eastern system greatly reduced the extreme risk-aversion of the Egyptian Empire and resulted in a more risk-neutral power-induced risk attitude. Consequently, the Egyptian civilization showed more energy and creativity after its inclusion within the Greater Near Eastern system.

The greater relativist concern and less extreme power-induced risk attitude generated better developmental achievements throughout the region. The second half of the second millennium (1500-1000 BC) was a time of great prosperity and progress for the Greater Near East. Advances were made in many areas. The manufacture of glass, for instance, was a major technological breakthrough. The first examples of glass vessels are found in northern Mesopotamia and dated back to the fifteenth century BC, whilst glazed bricks have been found in the palaces of the Middle Assyrian Kings.

Assyria was among the most prominent power players during this period. The Assyrian Empire was thoroughly based in a military tradition with every facet of the government, both civil and military, fitting the desired mold. Under the successive leadership of powerful kings, the Assyrian military numbered in the hundreds of thousands. This was an immense state capacity in the ancient

---

65 Refer to Wilkinson (2004).

world. Assyrian military prowess reached its zenith under Tiglat Pileser III. It was the Assyrians who first made broad use of iron, and thus established Assyria as the most technologically advanced power in the Near East. Enemies who relied on bronze weaponry could be dispatched with relative ease. It was the Neo-Assyrian Empire, with its powerful army equipped with iron weapons that ended the fragmentation of the ancient Greater Near Eastern international political system to establish the first Pan Middle Eastern Empire.

Assyrian achievements were not confined to the political and military spheres. Assyrian architectural, artistic and scientific achievements also reached their apex during this period. The irrigation system created by Sennacherib to redirect mountain spring water and the Khosser River to Nineveh, was first of its kind in the world. Assurbanipal was a scholarly individual who gained mastery of both Sumerian and Akkadian languages, and who could compute complex mathematical equations. Due to his academic interests, Asurbanipal assembled in Nineveh the first systematically collected and catalogued library in the ancient Middle East. Assyrian knowledge of the planets of our solar system led to accurate predictions of solar and lunar eclipses.[66]

The incorporation into the Greater Near Eastern international political system energized Egypt. Though Egypt had constant contacts with Middle Eastern Mesopotamian powers before this period, political-military interactions were insignificant and conquest by one over the other was almost inconceivable. With the involvement in the political military competition of the Greater Near Eastern system, the New Kingdom of Egypt adopted an expansionist policy of forward defense. It was the most powerful player in the Greater Near Eastern System. It conquered Nubia in the south and Syria-Palestine in the north. The main rival was the Hittite Empire. During this period, the concept of "warrior pharaoh" was introduced. Political-military competition with other major players in the Greater Near Eastern system propelled the ancient Egyptian civilization to reach its peak in the New Kingdom Era.

The New Kingdom Era was a period of nearly 500 years of political stability and economic prosperity and is considered the most golden of all epochs of ancient Egyptian history. Egyptian culture reached new heights during this period. A central government was created. This period produced an abundance of artistic masterpieces as testified by the rich store of treasures from the tomb of Tutankhamen (1347 – 1337 BC). The colonization of Nubia secured the largest source of gold in the ancient world. The wealth of the New Kingdom supported tremendous building activity and advances in art and architecture. Ramesseum, a temple to tell the greatness of Ramses II, had a library with 10,000 papyrus

---

66 Refer to Saggs (1991) and Haywood (1997, pp. 44-45).

scrolls. The period also produced the earliest historically attested expression of monotheism in the form of Atenism. The Egyptian economy was extremely rich. Hatshepsut, the third female pharaoh, encouraged trade, sending trade expeditions to Punt (present day Eritrea, Djibouti and northern Ethiopia). Thebes was one of the wealthiest cities in the ancient world.

Within this era of intense political-military competition in the Greater Near East, there were developments in state structure, bureaucracy, law, and market institutions, aiding the governance of the empires and coordination of effort across hundreds of miles. Imperial powers provided infrastructure such as road networks, which assisted traders. Complex arithmetical calculation appear for the first time on tablets from this age, a mathematical achievement to be surpassed only a thousand years later.[67] This Greater Near Eastern state system lasted from 1595 BC to around 700 BC. It survived until the arrival of the Neo-Assyrian Empire, the Neo-Babylonian Empire and then the Persian Achaemenid Empire: empires brought forth by a new military technology, the light cavalry.

5. Conclusions

Changes in military technology explain the history of the first civilizations. At the very beginning of written history in Mesopotamia and Egypt and before the invention of metallic weaponry, the economies of scale in warfare were low in both Mesopotamia and Egypt. The small mass factor prevented the consolidation of capability in the hands of the most efficient contestant. Consequently, the first civilizations had city-state systems during the earliest phase of their written history. The competition generated by the city-state systems of pre-dynastic Egypt and Mesopotamia was important for propelling civilization forward.

The introduction of bronze weapons increased the economies of scale in warfare and resulted in major structural changes in international politics of the first civilizations. Egypt was united early under one empire while the Mesopotamian city state system was soon transformed into a territorial state system and thereafter hegemonic powers and empires emerged. After the introduction of bronze weaponry, the complacency and conservatism of the Egyptian Empire which was a stable and almost perfect imperial order contrasted sharply with the creativity and energy of the pluralistic Mesopotamia.

---

67 Refer to McNeill (1999, pp. 35-37, 63-64).

The inventions of horse-drawn war chariots and iron weaponry introduced a more decisive and mobile form of warfare. Consequently, the isolated international political systems of Egypt and Mesopotamia were merged to become the Greater Near Eastern system, together with other adjacent civilizations, such as the Persians and Hittites. The formation of the Greater Near Eastern system due to the new military technology of horse-drawn war chariots and iron weaponry energized the lethargic Egyptian Empire. The political-military competition of the Greater Near Eastern system propelled the ancient civilizations to new heights of achievements.

In sum, the history of the first civilizations fits in well with the main arguments of this book. Major changes in military technology caused structural changes international politics. Such technological and structural changes have great impact on the relativist concern, power-induced risk attitude and economic performance of the constituent states. Specifically, the history of the first civilizations agrees with the main argument that political-military competition propels creativity and civilization, while political and military monopoly breeds complacency and conservatism.

CHAPTER 4
Classical Pluralism

1. Introduction

This chapter studies how military technology affected the geopolitical structure and economic performance of the greater Near East, Greece, Ganges India, China and Carthage. During this period, the defining military technological changes were the use of iron in warfare and, the rise of light cavalry and the combined arms legions, made up especially of light cavalry and light infantry.

While the pluralistic Greek, Ganges Indian and Chinese systems of this period bear resemblance to the early modern European Westphalian system with their even distribution of capability among the major contestants and emphasis on balance of power, the Greater Near Eastern system and the Carthaginian system are sharp contrasts with their extremely high concentration of capability within the hands of the imperial powers.

2. Light Cavalry

The use of iron in production allowed agriculture to spread to heavily forested areas and expanded the domain of civilizations. Consequently, the Chinese, Ganges Indian and Greek civilizations emerged. From around 800 BC to 200 BC, China, Ganges India and Greece generated splendid cultural achievements. This is referred to as the classical golden era in human civilization, or the Axial Period. In "Navigating world history," Manning (2003) comments on Jaspers (1949):

> "He defined "the Axial Period" (centering on 500 B.C.E. and including the three centuries on either side of that date) as a time of breakthrough in knowledge and faith, with insights that were never to be exceeded until the present day. This was the time of Confucius and Lao Tzu in China, of Buddha and the authors of the Upanishads in India, of Elijah and Jeremiah in Israel, and of Homer, Plato, and Thucydides in Greece. Jaspers sought the cause of the axial moment that he observed: he rejected the assertion of Alfred Weber that it had resulted from the influences of Indo-European horsemen dispersing across Eurasia, and concluded instead that it resulted from 'many small States and small towns: a politically divided age engaged in incessant conflicts; ... questioning the previously existing conditions.'" (51)

Jaspers (1949)'s conclusion on the cause of the splendid achievements of the Axial Period agrees with the main arguments of this book.

During the same period, however, the Greater Near Eastern region—the early leader in human civilization for more than two millennia—stagnated and fell behind. Around this period, the use of iron weapons and horses, first in horse-drawn war chariots and then in light cavalry units in warfare, raised the economies of scale in conflict for the Eurasian world. These military technological innovations facilitated the rise of combined arms legions with light infantry and light cavalry supplying the bulk of the fighting force, thus increasing the mass factor in warfare.[68] Warfare became more mobile, decisive and destructive. The core of the Assyrian army had been war chariots, but by the time of the Neo-Assyrian Empire, light cavalry had replaced the chariots. Wars with the Persians, who used light cavalry, had prompted the Neo-Assyrian Empire to reform its military and to rely more on light cavalry instead of the more cumbersome horse-drawn war chariots.

The more mobile, decisive and destructive form of warfare caused by the use of light cavalry and combined arms legions composed of light infantry enlarged the mass factor and caused greater disparity in the distribution of resources and capability within the Greater Near Eastern world. This new form of warfare ended the comfortable security of Egypt in international politics. From around 1000 BC onwards, foreign invasions of Egypt were more probable and frequent. In 712 BC the Nubians invaded Egypt to establish the 25th dynasty. They were supplanted by the Assyrians, who occupied Egypt between 671-651 BC. In around 605 BC, the Neo-Babylonians then defeated and captured Egypt, before the Achaemenid Persians took their turn with a further conquest of Egypt in 525 BC.

The Persians, who pioneered the use of light cavalry, finally conquered Mesopotamia and Egypt during their Achaemenid Dynasty. Persian warriors formed the core of the Achaemenid standing army. The army relied heavily on light infantry and light cavalry units, although there were also a small number of obsolete war chariots, fielded by the aristocrats. Occasionally, there were small numbers of camel-borne troops and war elephants present as well. The light cavalry was the most important branch of the Achaemenid army, instrumental in conquering far-flung subject lands and remained important until the last days of the Achaemenid Empire.

In China, Greece and Ganges India, the increase in the economies of scale in warfare associated with these military technological changes intensified competition within these diverse international political systems without destroying

---

68 Refer to Dudley (1992).

the city-state or territorial state systems. The distribution of capability within these systems remained highly even within this period. Consequently, in the Chinese, Greek and Indic systems, the positive scaling effect of a larger mass factor dominated the negative unbalancing effect and the relativist concern increased. Furthermore, the largely even distribution of capability within the Chinese, Greek and Indian systems prevented the emergence of extremely risk-averse or risk-seeking power-induced risk attitudes with their associated distortions. Therefore, within these systems, economic and developmental performance improved.

In the Greater Near East, these military technological changes had very different effects on geopolitics and economic performance. In the Greater Near East, the greater economies of scale in warfare caused a concentration of resources and military capability in the hands of the imperial regimes of the Neo-Assyrian Empire, the Neo-Babylonian Empire, and the Achaemenid Persian Empire. The increase in the asymmetry of relative capability reduced relativist concerns. The increase in the economies of scale in warfare caused a further reduction in relativist concern. The great asymmetry in distribution of capability within the Greater Near Eastern system also generated an extremely risk-averse power-induced risk attitude. Consequently, in the Greater Near East, creativity and economic performance faltered.

In the Western Mediterranean basin, besides the above military technological changes, the rise of gigantic naval fleets due to the use of triremes also increased the economies of scale in warfare. There was a high concentration of resources and capability in the hands of Carthage, a great maritime power strategically located at present-day Tunisia. Consequently, there was also a low relativist concern and a highly risk-averse power-induced risk attitude there and therefore a low level of innovation and developmental effort.

In sum, the same military technological changes had very different effects on the international political structure and developmental performance of the classical civilizations of China, Greece, India, Carthage, and the Greater Near East. The following sections will analyze each civilization in detail and will especially compare classical Greece with the Carthaginian Empire to study the effect of international political structure on developmental performance.

3. The Middle East

From the dawn of the first civilization to around the establishment of the Achaemenid Persian Empire in 525 BC, Middle Eastern leadership in human civilization had been unchallenged. The Near East had always been the most advanced region in the world for over three millennia. The Near East bestowed upon the world the first writing system, the first city and the first law codes, to name just a few. But during the Axial Period the Near East was overtaken by the classical civilizations of Greece, Ganges India and China as the pacesetters in human civilization.

During this period when leadership in human civilization shifted, there were a series of pan-Middle Eastern empires that conquered the whole of the Near East, Egypt and beyond. First was the Neo-Assyrian Empire, followed by the Neo-Babylonian Empire and the Achaemenid Persian Empire. The Neo-Assyrian Empire briefly occupied Egypt, from 671-651 BC. The Neo-Babylonian Empire controlled Mesopotamia, Egypt and parts of Persia, though its reign was short, from 604 BC to 562 BC. In 539 BC, Babylon itself was conquered by the Persians, and the Achaemenid Persian Empire soon became the largest of the pan-Near Eastern empires and the largest empire thus far in history. In sum, from around 700 BC, the Middle East was effectively governed by a series of universal empires.

Military technological changes drove the aforementioned structural changes in the Greater Near Eastern international political system. The use of iron weapons and the rise of light cavalry led to the emergence of combined arms legions where light infantry supplied the bulk of the fighting force and the light cavalry was the important mobile striking force. There was an increase in economies of scale in warfare associated with these military technological innovations as warfare became more deadly and mobile.

In the Greater Near East, there was a dominant core area in the form of the Fertile Crescent at the center, with the Nile Valley on the west and the western part of the Iranian Plateau on the east. Consequently, there was substantial asymmetry in combined relative economic and military efficiency in favor of the power controlling the dominant core area. The combined effect of the greater economies of scale in warfare and the more asymmetric combined relative economic and military efficiency generated a very high asymmetry in the distribution of capability and resources. The military technological changes therefore transformed the Greater Near Eastern system from a multi-polar pluralistic international political system into an imperial order with resources and military capability highly concentrated in the hands of the imperial authority.

The largest of these pan-Greater Near Eastern empires, the Achaemenid Persian Empire, actually expanded until reaching its natural boundaries. Several ancient seats of Middle Eastern civilizations, including Mesopotamia, the Indus River Valley, and the dry arid landscape of Egypt, were united under its rule. These were regions that were highly advanced economically and densely populated with terrains suitable for conquests by the Persian light cavalry and light infantry legions. Beyond the empire, to the east were the dense forests of the verdant Ganges river valley of India; to the west were the Mediterranean

Sea and the mountainous maritime geography of Greece; to the south were the Arabian and Sahara deserts; and to the north was the sparsely populated Eurasian Steppe. These were regions that either had terrains unsuitable for the military operations of the Persian light cavalry and light infantry legions or were unimportant economically and therefore not worth the effort of the empire to acquire. The empire reigned from 557-300 BC and there was no serious challenger to its supremacy.

Many will read "no serious challenger" and immediately think of the Spartans and other Greeks at Thermopylae. The victories of the Greeks over the Persians in the Greco-Persian Wars (499-449 BC) are indeed much recounted and celebrated in the Western tradition, for upon them hinged the survival and continuity of Western civilization. Yet, these battles were of no central concern to the Persians. These defeats undermined neither the rule of the empire nor its proper function: the rule of the Achaemenid Empire over the oriental provinces of Persia, Mesopotamia, Anatolia, Syria and Egypt was never threatened. These provinces were of much greater significance (than any small Greek city-state) to the health of the Persian Empire. Pastoral nomads of Central Asia, including the Scythians who resided around the Aral Sea, were persistent intruders. But they were merely irritants. They posed no serious threat to the wellbeing and survival of the all-powerful empire. Successfully presenting itself as vicegerent of the supreme deity Ahura Mazda on earth, the Achaemenid regime was almost uncontested from within as well as without. Revolts were rare and the imperial family's dominance was never challenged until the invasion by Alexander the Great.

The increased size of the mass factor generated extreme asymmetry in the distribution of resources and capability within the Greater Near Eastern international political system. The aggravated asymmetry in distribution of capability blunted political and military competition within the system, and lowered relativist concern. Moreover, given the highly concentrated distribution of capability induced by the military technological changes, the enlarged mass factor itself further dampened political-military competition and lowered relativist concern. Consequently, the series of pan-Middle Eastern empires, including and culminating in the Achaemenid Empire, had a very low relativist concern, given the almost nonexistent international political-military competition. Inevitably, development in economic and cultural fronts suffered. The great asymmetry in distribution of capability within the Greater Near Eastern system caused not just complacency in developmental efforts. It also generated an extremely risk-averse power-induced risk attitude, given the increased size of the mass factor. Consequently, in the Greater Near East, imperial complacency and conservatism prevailed and creativity dissipated.

This era of the pan-Middle Eastern empires of the Neo-Assyrian, Neo-Babylonian and Achaemenid Persian Empires, covering the period of roughly

700-300 BC, sees imperial regimes promoting the diffusion and mixing of cultures and effectively ended the isolation of the old civilizations. Yet, the weight of the successive empires stalled the developmental momentum in these old seats of civilization and the economy of the region slowly declined. Furthermore, in sharp contrast to the bursts of innovation found in the earlier era of a more pluralistic international political structure, the Greater Near Eastern world could boast of few major advances during this era of inertia.[69]

On the other hand, the adjacent areas of mountainous maritime Greece and the verdant Indian Ganges valley, as well as the more distant South India and even more remote China, were developing under their respective city-state and competitive territorial state systems and were bringing forth their classical civilizations with the representative figures of Socrates, the Buddha and Confucius.[70] These late-starter civilizations soon began to catch up with and even surpass the Middle East, whose ancient leadership in civilization disappeared. This happened despite a universal peace and other conveniences accorded by the Neo-Assyrian and the Achaemenid Persian imperial orders.[71]

4. Greece and Carthage

Greece is highly fragmented geographically. Greece has about 3,000 islands, including the famous islands of Crete and Rhodes. The mainland contains multiple mountain ranges that split the country into many peninsulas and isolated regions, including the Peloponnesus Peninsula. The fragmented geography encouraged localism in politics during antiquity. During the classical era, the dominant political units in Greece were the poleis – independent city-states of which the more famous and important include Athens, Sparta, Corinth and Thebes.

Another important geographical feature of Greece is its intimacy with the sea. Socrates the Greek philosopher commented that "We live around the sea, like frogs around a pond." Greece has an extended coastline of over 15,000 kilometers; no point on the Greek mainland is farther than 100 kilometers from the sea. Due to the close proximity to the sea, the classical Greeks were a seagoing people with extensive trading networks and colonies in the Black Sea and western Mediterranean regions. The existence of extensive and far-flung Greek overseas colonies which were politically independent from their mother

---

69 Refer to Wesson (1967, 1978).
70 Refer to Haywood (1997, pp. 22-23).
71 Refer to Stavrianos (1982, Part II).

cities made concentration of capability even harder and caused further fragmentation in Greek politics.

After slowly recovering from the Greek Dark Ages of ca 1200-800 BC, city-states began to emerge in Greece around 900-800 BC. In addition, around 800 BC the main period of Greek overseas expansion also began. The use of iron implements facilitated the spread of agriculture into harsher terrain, and the advances in agriculture aided the resurgence and extension of the civilization.

The use of iron spread into warfare and created an infantry revolution around 700-650 BC, which saw the rise of the phalanxes or the heavy infantry legions. A Greek phalanx was a close formation of heavily armored infantrymen specialized in prolonged combat. These infantrymen were named hoplites after their shield *Hoplons* (a large circular shield covered with a bronze sheeting), the hoplites were armored typically in a leather cuirass (although some wore a bronze cuirass or no armor at all), the *Corinthian* helmet further supplemented by greaves. The weapon of a hoplite was a weighted spear called the *Dory*. The phalanx formation organized these heavily armored infantry into a bronze shield wall, which during this period was the dominant infantry doctrine. Sparta took the lead in developing the phalanx through its rigorous training regimes creating a fully professional force of hoplite soldiers. Sparta dominated the Peloponnesian Peninsula around 650 BC, at a time when written law codes were being created. This infantry domination of warfare covers the period ca 500 BC-300 AD, which roughly (and not coincidentally) corresponds to the time from which the ancient Greeks began to unify their city-states and conduct large-scale battles, to the dissolution of the Western Roman Empire 800 years later.

Greek phalanxes were renowned for their mass and might throughout the Near East and the Mediterranean region. Consequently, many Greek soldiers served as mercenaries for Persian governors. The military technological changes increased the economies of scale in warfare. Politics and wars increasingly involved larger alliance systems and empires. Given the fragmented mountainous and maritime terrain, mounted soldiers initially played a relatively minor role in Greek warfare (this later changed under the leadership of Philip II of Macedon who pioneered combined arms of cavalry and infantry formations). Military tactics in the Greek city-state system era took advantage of infantry mass envelopment and flanking maneuvers.

The greater economies of scale in warfare favored larger political units. Relatively large states or empires, encompassing stretches of settled civilized societies with their dense population and high level of material wealth, have the benefit of being able to field large numbers of battle-efficient legions. There was within Greece, therefore, a trend towards ever larger political organization in either the form of military alliances or large territorial states or empires. This trend toward larger political units culminated in the Macedonian dominance of Greek politics from ca 350 BC.

Macedonia, under King Phillip and Alexander the Great, pioneered the use of combined arms warfare with heavy infantry at its core. Phillip increased the training and importance of cavalry on the battlefield, but the cavalry was to work in tandem with the Macedonian phalanx. Phillip of Macedonia further supplemented the destructive power of the Greek heavy infantry by incorporating missile auxiliaries, well trained light infantry (hypaspists) and siege equipment. The combination of the revolutionized Macedonian phalanx, strong cavalry formation, elite light infantry supplemented further with siege corps resulted in more decisive battles and campaigns. The organizational and tactical innovations of Phillip of Macedonia were fully and boldly exploited by his son Alexander the Great first to subdue the Greek city-states and then to take on the Achaemenid Persian Empire. The lethargic Persian Empire had been too thoroughly plagued by imperial complacency and conservatism and was not in a position to react decisively and successfully to the sudden emergence of threat due to these military innovations. It was swiftly overrun and conquered by Alexander the Great.

There was a military revolution in the navy as well. The use of triremes resulted in the formation of specialized and unprecedented war fleets. Economies of scale in naval warfare became larger, making naval warfare more expensive. The greater financial requirement contributed to the consolidation of city-states into large alliances and empires, and Athens became the premium Greek naval power.[72] These two military revolutions, on land and sea, enabled the Greek city-states to defeat the Persian Empire in the Greco-Persian Wars: the Persian light infantry were no match for the densely packed and heavily armored Greek phalanxes. These military revolutions also began a trend towards larger political organizations in the Mediterranean basin, the apex of which was the Roman Empire.

The polis as a form of political organization signified the maturing of Greek societies and the ability to tap and mobilize resources for increases in scale in warfare, land or sea. Yet, creating a navy, manning its vessels, and replacing worn-out warships at least every 20 years required the resources and manpower of much larger communities than Athens, Corinth, or Sparta. Therefore, some form of pooling of interstate resources was almost essential for any Greek strength at sea. The Greek city-states ultimately proved too small to exploit the greater economies of scale in warfare generated by the new military technology. Consequently, the fragmented classical Greece city-state system was replaced by the empire of Macedonia.

During the long existence of the Greek city-state system, the isolated and fragmented geography of Greece caused the city-states to be fairly equal in

---

72 Refer to Starr (1989, pp. 50-66).

terms of relative combined military and economic efficiency. In other words, the classical Greek international political system was pluralistic and had a highly even distribution of resources and capability. Political-military competition was a constant fact of life in the classical Greek system and relativist concern was high.

The military revolution on land and at sea resulted in an increase in the size of the mass factor. Yet, given the highly symmetrical relative combined military and economic efficiency and very even distribution of resources and capability, the enlarged mass factor generated basically no greater asymmetry in distribution of capability and produced hardly any negative unbalancing effect and almost only a positive scaling effect. Relativist concern was significantly raised. The military revolution of heavy infantry and large war fleets heightened political-military contests among the Greek city-states and prompted them to greater achievements and state power. Furthermore, the highly even distribution of capability in the Greek system ensured that a largely moderate power induced risk attitude prevailed. There was no serious distortion in economic decisions or allocation of resources caused by severe risk-seeking or risk-averse power-induced risk attitudes. Consequently, the Greek international political-military environment was highly salubrious to creativity and development.

The most outstanding achiever of classical Greece was Athens. Despite being almost constantly at war, like other Greek city-states, Athens flourished culturally, economically and politically in the 5th century BC. Athens dominated the Eastern Mediterranean and Black Sea trade and owned silver mines. Buzan and Little (2000, p. 200) note that: "Athens, for example, imposed a single currency and standardized weights and measures on the other city-states within its empire (Meijer and Van Nijf 1992: 33-51)." Democracy developed alongside an exceptional outburst of cultural confidence and creativity, in Greece as a whole and Athens in particular. Vase-painting, sculpture and drama reached new heights. A great public building program began: the city, and notably the Acropolis were rebuilt and the classical style of art and architecture matured.[73] Many of the most important authors of Western culture were products of this period. These include the dramatists Aristophanes and Sophocles; the philosophers Socrates, Plato and Aristotle; and the historians Herodotus and Thucydides.

Rivalry among the Greeks and the completely Hellenized Macedonia propelled Greek achievements in the economic, cultural, political and military arenas to new heights.[74] Herodotus completed his *History* in 430 BC; Pheidias

---

73 Refer to Haywood (1997, pp. 82-83).
74 Refer to Bernholz (1998, pp. 109-114) in Bernholz et al. (1998) and Ungern-Sternberg (1998, pp. 85-107).

and Polyclitos produced their sculptures and brought art to new heights; Democritos developed an atomic theory of matter; Thucydides wrote *The History of the Peloponnesian War*; Socrates taught his philosophy; and Plato and Aristotle founded their schools, to name but a few of the cultural achievements of the time.

After the Macedonian conquest of the Persian Empire under Alexander the Great, Greek cultural influence spread throughout the classical world. Classical Greek became the lingua franca. Political and military competition of the pluralistic Hellenistic international political system further propelled the development of the classical era. (Remember that, after the death of Alexander, his empire divided and the pluralist Hellenistic system emerged from it.) The Hellenistic Age (336-31 BC) generated extraordinary economic and scientific achievements.[75] Alexandria, the capital of Ptolemaic Egypt, became the largest and richest city in the world, and came to overshadow Athens as a cultural center. Around 105 BC, a college of technology was founded at Alexandria. Science made important advances: Aristarchus first proposed a heliocentric universe around 270 BC; Eratosthenes measured the Earth's circumference around 235 BC; Euclid of Alexandria wrote the famous "Elements" on geometry which is one of the most influential works in the history of mathematics and remained the main textbook until at least the late 19th century. These achievements, under first the Greek city-state system and then the Hellenistic state system, were collectively known as the Greek Miracle.

Stavrianos (1982) comments on the relationship between the international political competition and the technological achievements of the classical world:

> "Significantly enough, it was only war that could arouse the classical civilizations from their technological lethargy. The Greeks invented ingenious ratchet-equipped catapults and wheeled assault towers pulled by block and tackle." (184)

The comment of Stavrianos (1982) agrees with the main argument of this book that political military competition propels creativity and economic development.

The splendid achievements of classical Greece contrasted sharply with the unmemorable performance of the nonetheless mighty Carthaginian Empire. The Carthaginian Empire flourished from around 800-264 BC (the year of the First Punic War). Carthage, a Phoenician colony, had been established

---

75 Refer to Stavrianos (1982, pp. 98-102).

ca 814 BC. 800-600 BC was the main period of Phoenician and Greek colonization in the western Mediterranean region. The Carthaginian system had the upper hand in the western Mediterranean region: around 580 BC the Carthaginian Empire defeated Greek settlers at Lilybaeum (Sicily), and in 539 BC an Etruscan-Carthaginian force expelled the Greeks from Corsica.

The Carthaginian Empire was the first true maritime empire in ancient times. By 500 BC the Carthaginians were the dominant force at sea in the Western Mediterranean, whilst the Persians dominated the eastern waters.[76] The Carthaginian Empire largely dominated the seas and coastal areas of the western Mediterranean region until the Punic wars. As a commercial naval empire, the Carthaginian Empire had practically expanded to her natural boundaries. Encompassing Corsica, Sardinia, the Balearic Islands, Morocco, Algeria, Tunisia, the western part of Sicily, the southeastern part of Spain, and the western part of Libya, Carthage had no worthwhile rival in the Western Mediterranean basin. The Carthaginian Empire's economic might depended on trade. The height of its dominance in the Western Mediterranean occurred between 600-500 BC. At the peak of its power, Carthage ruled over 300 other cities around the western Mediterranean basin and was the leader of the Phoenician world.

In the Western Mediterranean basin, Carthage reigned supreme. Carthaginian naval mastery in the Western Mediterranean world was unquestioned until the third century BC, but this power was in fact the cause of its mediocre developmental achievements.[77] Toutain, in his book *The Economic Life of the Ancient World (Economie Antique)* (1930) affirms Carthaginian trading and agricultural achievements but comments that Carthage only achieved economic success, with no concomitant political, intellectual or ethical achievements. Given its secure external environment, the Carthaginian Empire was a loose, superficial network of business and naval bases and maintained no centralized state machinery.

The Carthaginian Empire was very secure because of the almost total absence of political-military competition—until the Punic Wars. Relativist concern was low and there was a lack of drive for greater development. Both the naval revolution associated with the use of triremes and the heavy infantry revolution associated with the rise of phalanxes increased the mass factor. Given the extreme asymmetry in distribution of resources and capability in the Western Mediterranean basin, the enlarged mass factor furthered dampened political-military competition in the Carthaginian system and lowered relativist concern, since the negative unbalancing effect dominated the positive scaling

---

[76] Refer to Starr (1989, pp. 25-28).
[77] Refer to Starr (1989, p. 54).

effect. Moreover, the high concentration of capability and the large mass factor generated a highly risk-averse power-induced risk attitude which severely distorted economic decisions and especially inhibited creativity.

A comparison between the Carthaginian Empire and the Athenian Empire reveals that although the Carthaginian Empire ruled much longer and was more powerful, its memorable achievements pale in comparison to those of the Athenians, who were driven by the competition from other Greek city-states to excel and achieve. Pericles, in his famous funeral oration commemorating the Athenian soldiers who had fallen in battle against Sparta in 431 BC, exclaimed: "to sum up: I say that Athens is the school of Hellas......"[78] If one contrasts the geographical sphere of influence of the Greeks with that of the Carthaginians, then one finds that the Carthaginian Empire had about the same sphere of influence from the Achaemenid Persian era through to the time of the first Punic war, while the Greek city-state system expanded to almost the whole of the classical world during the Hellenistic Era. This is yet another piece of evidence for the stark difference in achievements and development drive between pluralistic Greece and imperial Carthage. A simple list of famous names can perhaps sum up the difference: asked to name famous Carthaginians, few could probably remember more than Hannibal; but how many ancient Greeks could be named?

5. India

The earliest civilization of India was the Bronze Age civilization of the Indus River valley. It collapsed around the beginning of the second millennium BC and never recovered. Little is known about the history of this civilization. Later, from ca 1500 BC to 500 BC, came the Vedic period in Indian history. This era saw the spread of Indo-Aryan civilization across the whole of northern India, especially the Gangetic Plain. City-states began to emerge around 1000 BC in the Ganges River valley and also southern India. By 900 BC small tribal kingdoms and aristocratic tribal republics, known collectively as janapadas, were developing across the Ganges plain. By 700 BC, by means of agrarian extension, control of trade routes and a new and more aggressive type of warfare partly caused by the use of iron weapons, they had consolidated to form sixteen mahajanapadas (that is, great realms).[79] From 540-490 BC, Magadha, under its energetic king Bimbisara, had emerged as the most powerful mahajanapada.

---

78 Refer to Jowett (1999, pp. 35-46).
79 Refer to Kulke and Rothermund (1998, pp. 33-49).

The Indo-Aryan civilization centered on the Ganges River valley. The Ganges River has many tributaries, and its basin covers an extensive area in northern India. Unlike the arid Indus river valley, the Ganges region has high precipitation rates. Densely forested during ancient times, the harsh terrain of the Ganges Basin was a difficult barrier for ambitious conquerors. The extreme humidity and heat of the Indian summer deterred foreign conquerors like Alexander the Great, Genghis Khan and Timurlane. The hostile terrain helped in the maintenance of the initial city-state system and subsequently the territorial state system. Indian political geography was therefore highly fragmented during the formative period of the Ganges Indic civilization.

Another important region of the subcontinent is the mountainous southern India. Southern India is dominated by the Deccan Plateau. In southern India, mountain ranges and rivers running west to east break up the region into many isolated compartmentalized localities; political unification was even harder to achieve and maintain in southern India than it was in the north. The existence of highly fragmented southern India meant that even if an imperial power controlled the Indo-Gangetic Plain, it was still difficult for the imperial regime to control peninsular southern India.[80] Consequently, it was more difficult for an imperial order that covered the whole civilization to emerge and survive in India than it was in the Middle East or China.

Given the fragmented geography, with the exception of the brief unification under the Mauryan Empire, the Indian subcontinent was fragmented politically throughout antiquity—indeed, the unification of the subcontinent is historically quite a recent phenomenon. The absence of a dominant core area meant that relative combined military and economic efficiency among the many contestants was fairly symmetric. The fragmented geography also meant that with the same military technology, the mass factor was smaller in India than in China or the Middle East. A fairly symmetrical relative combined military and economic efficiency and a small mass factor resulted in a fairly even distribution of resources and capability among the many political-military contestants of the Indian system. The fairly even distribution of capability also meant that there was no extremely risk-averse or risk-seeking power-induced risk attitude to severely distort economic decisions. Consequently, the high relativist concern of this period in India generated good developmental performance.

During this period there was widespread commercial development, political and cultural reforms, as well as religious reformation and counter-reformation.[81] Hand in hand with the process of state formation was the growth of

---

80 Refer to Kulke and Rothermund (1998, pp. 9-12).
81 Refer to Stavrianos (1982).

cities and great development of religions. This was the formative period of the Hindu religion. The late 6th century BC also witnessed the lives and teachings of Mahavira, the founder of Jainism and of Siddhartha Gautama, the Buddha. There was geographical expansion of the Gangetic Indic civilization as well. By 500 BC the Indic civilization extended as far south as the Godavari River, covering the whole of northern India.[82] There was use of standardized weights and emergence of scholars such as the great linguist, Panini.[83]

The application of iron weaponry from the 7th century BC increased the scale of warfare and led to the use of large infantry legions; around 500 BC elephants were introduced into warfare alongside massive light infantry legions. A strong fiscal capacity of the state was required to finance the larger and more expensive war machine.[84] On the size of the military, Kulke and Rothermund (1998) notes that:

"Greek and Roman authors report that the Nandas, who had their capital at Pataliputra when Alexander the Great conquered northwestern India, had a powerful standing army of 200,000 infantrymen, 20,000 horsemen, 2,000 chariots drawn by four horses each, and 3,000 elephants. This is the first reference to the large-scale use of elephants in warfare. Such war elephants remained for a long time the most powerful strategic weapons of Indian rulers until the Central Asian conquerors of the medieval period introduced the new method of the large-scale deployment of cavalry."(56)

The new form of warfare had a large mass factor. The increase in the size of the mass factor generated more intense political-military competition. Given that there was a fairly even distribution of capability, the larger mass factor generated a positive scaling effect that dominated the negative unbalancing effect. The consequence was to raise relativist concern and to further propel development in the Indian subcontinent. Moreover, there was no extreme risk-averse or risk-seeking power-induced risk attitude, given the fairly even distribution of capability. The heightened relativist concern was therefore translated into creativity and greater developmental effort without much distortion.

These military innovations increased the economies of warfare and spurred the consolidation of city-states into the mahajanapadas. Diffusion of knowledge of Persian imperial administration helped to increase asymmetry in relative combined economic and military efficiency in favor of the strongest contestant. The enlarged mass factor, together with the aggravated asymmetry in

---

82 Refer to Haywood (1997, pp. 100-101).
83 Refer to Kulke and Rothermund (1998, pp. 50-53).
84 Refer to Kulke and Rothermund (1998, pp. 55-57).

relative combined military economic efficiency, generated the short-lived uni-fication under the Mauryan Empire, when exceptionally ambitious and capable leaders emerged in the scene.

The Mauryan universal empire ruled briefly from 322-185 BC as a central-ized state.[85] Ashoka, a Mauryan Emperor, imposed the Buddhist doctrine of right conduct throughout the empire. That period also saw the beginning of the Hel-lenistic Era, brought in by Alexander and his conquests. The work of Chanakya Kautilya, the *Arthashastra*, was the culmination of the art of statecraft of the an-cient Indian civilization. Mauryan imperial rule however was extremely short. It effectively lasted only three generations- from Chandragupta to son Bindusara and grandson Ashoka. The empire collapsed soon after the death of Ashoka. So, the developmental drive of the pre-Mauryan pluralistic international political order was not choked off. The momentum was retained and reinforced during the subsequent period to create the classical golden age of India.

6. China

The recorded history of the Chinese civilization begins with the Shang Dynas-ty, which reigned from around 1600-1066 BC as the first urbanized culture of China. (The earlier Xia Dynasty, 2100 to 1600 BC, has no recorded history; what historians know about the Xia is mainly through mythology.) The pow-er of the Shang Dynasty was quite limited as it covered only the middle and lower stretches of the Yellow River valley: there were many rival states and tribes both within its territory and beyond its border. The Shang Dynasty was overthrown by a tribal alliance from the west around 1066 BC, and that tribal alliance installed the Zhou Dynasty.

The territory or sphere of influence of the Zhou Dynasty was quite limited as well and was only slightly more extensive than that of the Shang Dynasty. For example, the Kingdom of Chu in the valley of the Han River (a tributary of the Yangtze River) did not yield to the authority of the Zhou Dynasty, and the Zhou regime was unable to alter the situation. There were also many indepen-dent non-Chinese states or tribes yet to be pacified within the sphere of influ-ence of the Zhou Dynasty. The Chinese system of the time was held together more by common culture, blood ties and identity, rather than through political organization or military might. Governance was based on personal, familial

---

85 Refer to Kulke and Rothermund (1998, pp. 58-59).

or tribal ties - the ideal politics of Confucius. In many parts of ancient China, tribes rather than states were the dominant form of political organization.

The history of the Zhou Dynasty can be divided into two eras: the Western Zhou Dynasty (1066-771 BC) and the Eastern Zhou Dynasty (770-256 BC). By the time of the Eastern Zhou Dynasty, the growing power of the larger peripheral vassal states of Zhou had already overshadowed the authority of the central government. In turn the reign of the Eastern Zhou Dynasty can be subdivided into two eras: the Spring and Autumn Era, and the Era of the Warring States.

During the Spring and Autumn Era, military rivalries between the vassal lords became more intense and the authority of the Zhou Dynasty dwindled to almost nothing. This era was also known as the Era of the Five Hegemons. Wars became more frequent and were of larger scale, with smaller states annexed on a frequent basis. The number of states declined as reorganization occurred to wipe out smaller and weaker states. At the beginning of the Spring and Autumn Era, there were more than a hundred principalities. By the Warring States Era, this had dwindled to about a dozen, seven of which were the major powers.

The intense political-military competition generated high relativist concern. Furthermore, given the fairly even distribution of capability, there was no extremely risk-averse or risk-seeking power-induced risk attitude to distort economic decisions. Consequently, there were many reform efforts in the cultural, economic and social arenas, all seeking to harness greater power for the state in international political-military competition. The famous statesman, Guan Zi, for instance, was a product of the Spring and Autumn Era. Guan Zi (725-645 BC), who is also referred to as Guan Zhong, was a merchant who entered politics to become the prime minister of Duke Huan of the Dukedom of Qi. His reforms, many of which were mercantilist in nature, propelled the Dukedom of Qi to hegemony. Similar reforms were undertaken in the other states, especially the subsequent hegemonic powers.

China had a very pluralistic and competitive international order during the Era of the Five Hegemons and the succeeding Era of the Warring States. Given the highly even distribution of capability, there was no extreme risk-averse power-induced risk attitude to hold back the creativity or to distort the economic decisions of the rival states which were eager to increase their power. Consequently, the harmony offered by a common culture and identity of the earlier period broke down and traditional beliefs were questioned. Many philosophic schools blossomed. The phenomenon was referred to as "the hundred schools of philosophy". This was an era of great cultural, economic and political experiments and achievements which laid down the foundations of the Chinese civilization and state.[86]

---

[86] Refer to Hui (2005) for a comparison between the classical era Chinese competitive state system and the early modern European competitive state system.

The subsequent Era of the Warring States saw a great increase in large-scale warfare, the emergence of larger states, and a more centralized form of government with huge professional bureaucracy. The increase in the economies of scale in warfare was due to the rise of massive light infantry legions and light cavalry, in place of the horse-drawn war chariots used by the early Zhou Dynasty. Selective breeding had made possible more-powerful horses that could carry riders, rather than just pull chariots; technological advances such as the use of iron weaponry and crossbows made massive infantry legions the dominant fighting force on the battleground. The result was a form of warfare that was more mobile, deadly and decisive.

Given that the Chinese international order was highly pluralistic and relative combined economic and military efficiency was highly symmetrical, the distribution of resources and capability was fairly even among the major contestants. Consequently, the enlarged mass factor generated a mainly positive scaling effect which dominated the negative unbalancing effect. Political-military competitions among the contesting states were intensified. To be less efficient and powerful had dire consequences. There was therefore heightened relative concern which further pushed on development in many areas. Culturally and economically, this was a very productive and prosperous era. Political development during this period was spectacular too.

This period witnessed the further consolidation of the centralized bureaucratic states which had started in the Spring and Autumn Era. Every state undertook institutional changes in order to tap and utilize resources they controlled more efficiently, all so as to be able to compete in the international arena. The aristocracy lost their dominance of the military, which out of necessity became professionalized—as did the bureaucracy. The world-famous "The Art of War" by Sun Tzu was written during this era as were a myriad of other less famous works on military science.[87]

Intense interstate rivalry during this era prompted states to undertake reforms to advance their economy and to strengthen their state capacity. Many of the philosophy schools taught measures to enhance state strength. The Legalist school especially had this as its professed objective and advocated draconian measures to increase the economic, fiscal and military strength of the state.[88] A good indicator of the economic might and state capacity of the era was the length and scale of wars conducted. It was common for participating states to mobilize hundreds of thousands of soldiers, or even close to a million troops, and fight over extensive territory for months or even years. This

---

87 Refer to Sawyer (1993).
88 Refer to Fung (1966) and Bodde and Fung (1983).

could not have happened without a very high degree of economic achievement and fiscal strength.[89]

Many of the reform measures undertaken by the contesting states were pioneered and advocated by the already-mentioned Legalist school. The Legalist school of thought strongly advocated institutional reforms to enhance state capacity. The Legalists taught statecraft to princes and strongly supported the centralization of state administration. The basic institutional structure of the traditional Chinese state was laid down during this era, largely following the institutional structure of the Kingdom of Qin: the kingdom of the First Emperor who united China.

The Kingdom of Qin rose to military and political dominance after Lord Shang, a leading Legalist scholar, overhauled its institutions. The reforms undertook by Lord Shang made the state of Qin a powerful, centralized war machine. The new institutions installed included the centralized provincial-county system (replacing the feudal regime of the early Zhou Dynasty), the draconian penal code, and the professional bureaucratic and military organization of the central government.

During this period, one of the two most important developments in Chinese legal history took place: the writing of Fa Jing (Canon of Laws) by Li Kui of the Kingdom of Wei in 542 BC. (The other was the issuance of Kaikuang Lu (Kaihuang Code) by Yang Jian, the founding emperor of Sui Dynasty).[90] Although there were earliest publications and codifications of laws, such as those undertaken by premier Zi Chan of the Kingdom of Zheng during the Spring and Autumn Era, the work of Li Kui laid the basic framework for the legal institutions of China for the next eight to nine centuries. The legal systems of the Qin, Han and Jin dynasties basically followed the work of Li Kui.

Intellectual achievements were no less amazing. Almost all the major Chinese philosophic schools of thought had their origin in this period of great political turmoil and competition. The four major schools were Confucianism, which became official orthodoxy in imperial China, the statist Legalism whose teachings laid the foundation for the unification of China under the First Emperor, the pacific Moism and the naturalist philosophical Taoism.[91] Other minor schools included the Yin-yang school that theorized about cosmic forces and the basic elements of the universe, the logicians who focused on definition and logic, the school of agriculture that emphasized and promoted farming, the school of diplomacy that specialized in diplomatic politics and the school of

---

89 In contrast, the first war of such scale in the Western world was the Napoleonic wars of Europe.
90 Refer to Head and Wang (2005).
91 Refer to Fung (1966) and Bodde and Fung (1983).

military that studied warfare and strategy. There was such a burgeoning of intellectual inquiry that Chinese historians refer to the intellectual achievements of this era as the Hundred Schools of Philosophy.

Another piece of evidence testifying to the vigor of this era was the energetic geographic expansion of the Chinese sphere of influence. Almost all of the traditional areas occupied by the Han Chinese were explored and conquered during this era. The southern part of the Manchurian plain was conquered by the Kingdom of Yan. The Kingdom of Zhao pushed back the nomadic tribal alliance of Xiongnu and annexed Inner Mongolia. The Kingdom of Chu extended its administration all the way to what is Yunnan province today. The Kingdom of Qin acquired the Gansu and Sichuan provinces. The boundaries of the traditional Chinese state, as seen even today, were basically established at this time, with later periods seeing but minor expansions.

Economic achievements were startling. The era saw a huge volume of inter-state commerce. The merchant class achieved a high social status never to be repeated in later history; many merchants became statesmen, a move almost unimaginable in many of the later Chinese empires.[92] Urbanization advanced to a very high degree: a dozen cities had populations in the hundreds of thousands. There was mercantilist thinking and economic inquiry. The most famous of these merchant-statesmen was Lu Bu Wei (ca 290-235 BC), a merchant who entered politics to become the Prime Minister in the Kingdom of Qin.[93]

7. Conclusions

In sum, the broad situation across Eurasia during the early classical era fits in well with the main argument of this book that major changes in military technology caused structural changes in international politics and, such technological and structural changes have great impact on the relativist concern, power-induced risk attitude and economic performance of the constituent states. Specifically, the history of the early classical world agrees with the main argument that a pluralistic international order encourages creativity and development and a monolithic international order results in conservatism and stagnation. During the early classical era, the pluralistic and competitive international political systems of Greece, India and China propelled their civilizations ahead and created splendid classical achievements while the monolithic international

---

92 Refer to Wesson (1978, pp. 44-45).
93 For the historicity of Guan Zi, refer to Fung (1966) and Bodde and Fung (1983).

political systems of Greater Near East and western Mediterranean region stagnated. Building upon the achievements of the Near Eastern civilizations and propelled forward by the political military competition generated by the Greek city state system and Hellenistic state system, classical Greece was the most advanced of this era. Consequently, the ancient Near Eastern leadership in civilization ended.

CHAPTER 5

Gupta Efflorescence

1. Introduction

This chapter studies how military technology shaped geopolitics and economic performance of the late classical world. The battlefields of the late classical world were largely dominated by heavy infantry legions. During this era, massive continental-sized empires were the norm: namely the Chinese, Persian and Roman empires. The late classical Chinese, Persian and Roman systems are sharp contrasts to the early modern European Westphalian competitive state system: they had extremely high concentration of capability.

Late classical India is of special interest. The Hume-Kant hypothesis of Bernholz et al. (2004) is unable to successfully explain the cultural creativity and economic vitality of the Indian classical golden age. During the late classical era, India was under the reign of the powerful Gupta Empire. An international political order dominated by a powerful empire was not what the Hume-Kant hypothesis expects for a period of cultural creativity and economic achievements. In contrast, the argument proposed by this book has no difficulty in explaining the Indian classical golden age.

2. Heavy Infantry

The Greeks pioneered the development of heavy infantry. They packed fully armored infantry men into dense formations that charged forward like a tank—the so-called Greek phalanx. The Spartans were famous for being the masters of this tactic in battle. With this innovation, the tiny Greek city-states stood up to the gigantic Achaemenid Persian Empire. King Phillip of Macedon built upon the Greek success and established sophisticated combined arms within his armies with heavy infantry forming the core. The combination of light cavalry, missile weapons and siege apparatus increased military effectiveness and added economies of scale. Of special importance was the adding of light cavalry, which made possible deadly pursuit after the heavy infantry core had already delivered a decisive charge. King Phillip's son, Alexander the Great, capitalized upon the military revolutions introduced by his father to capture the whole of the Achaemenid Persian Empire.

In the early periods of the Roman republic the Romans utilized the Greek phalanx. Close-order heavy infantry was also the backbone of the Roman legions. Over time the Romans discarded phalanx as their primary strategy however did reserved the phalanx as a last resort weapon against enemy pursuits or charges. The phalanx could charge with great force, but only on flat plains, and was therefore a cumbersome military unit to maneuver in difficult and complicated terrain. The Romans developed the maniple formation which was more flexible and could more easily be adapted to different complicated terrains. Roman victories in the four Macedonian Wars (215 to 205 BC, 200 to 196 BC, 172 to 168 BC and 150 to 148 BC) against the much vaunted Macedonian and Greek phalanxes proved the battle value of the Roman military innovations.

With the emergence of heavy infantry combined arms legions, warfare entered a new era. There were greater economies of scale in conflict. Armies grew in size, and so did political units. In Europe, the tiny Greek city-states gave way to larger territorial Hellenic states which were, in turn, conquered by the continental-sized universal Roman Empire. This trend toward greater political agglomeration occurred in Middle East and South Asia as well. In the Mediterranean region, this increase in the economies of scale in warfare was reinforced by the parallel development of triremes and gigantic war fleets which aided the rise of the Roman Empire.

At the other end of Eurasia, in China, the military use of iron weapons started at the end of the Spring and Autumn Era though bronze weapons were still popular even in the Era of the Warring States. Besides the rise of massive infantry legions, the replacement of horse-drawn war chariots by light cavalry as the main quick mobile force at around that time increased military decisiveness too. The invention and use of crossbows further enhanced the rise of massive infantry legions and made warfare more destructive. It was these massive powerful infantry legions that made possible the unification of China under the First Emperor.

3. Roman, Persian and Chinese Empires

Greater economies of scale in warfare due to the rise of heavy infantry favored the Romans over their rivals. The Roman Republic, an agrarian society with a large pool of manpower, was able to field larger numbers of legions to overwhelm and outlast their opponents. The resolve of the Romans to endure great losses in war till final victory made the Roman Republic a formidable enemy. Before the rise of Rome, in the Eastern Mediterranean world, the rough balance of power between the three Hellenistic states of Macedonia, Ptolemaic Egypt and Seleucid Syria was maintained for approximately a century. However, Roman conquests proceeded fast from around 200 BC, after Rome had defeated the leading trading power of Carthage and its mercenary military in the Western Mediterranean region. The result was the Roman domination of the whole Mediterranean basin.

After defeating Carthage, the Roman navy had unchallenged control of the Mediterranean Sea. The control of the Mediterranean Sea further shifted relative combined military and economic advantage to the Romans, with further advantages in logistics and mobility. Roman fleets could quickly transport large quantities of food, equipment and troops to the campaigning regions. Aided by the control of the sea, the Roman heavy infantry legions were almost invincible in the Mediterranean Basin. After the conquest of the Carthaginian Empire, Rome proceeded to conquer the whole Mediterranean world and beyond; although the republican form of government ended in 48 BC, when Caesar defeated Pompey, the Roman conquests stopped only in 117 AD.

The larger mass factor created by the heavy infantry military revolution, and the more asymmetric relative combined military and economic efficiency (partly due to the Roman control of the sea) combined to generate a very high concentration of resources and capability in the Mediterranean region. By the late classical era, the whole Mediterranean world was in the firm grip of the Roman imperial order. Beyond the borders were tribes and nomads who made incursions but posed no organized challenge to Roman authority. Worthy rivals to Roman power were no real threat: the Chinese empire of the Han Dynasty was at the other end of the Eurasian landmass, and therefore too far away; the Persians were closer, but occupied a very different kind of geopolitical niche, that of desert, steppe, highland and oasis, and therefore were fighting with the Romans only over margin lands. The Roman Empire had extended to its natural boundaries: to the south was the Sahara desert, to the west was the Atlantic Ocean, to the East were the deserts, steppes, highlands and oases of the Persian Empire and, to the north were the forests of the Germanic tribes. The Roman imperial order survived for over five centuries or more, practically uncontestable within its own geopolitical niches.

The high concentration of capability in the hands of the Roman Empire lowered relativist concern. Political-military competition was further dampened due to the large mass factor and Roman supremacy, and relativist concern took a further dip. Consequently, in the Roman lands, stagnation and decline resulted. Moreover, the combination of a high concentration of capability and a large mass factor brought forth an extremely risk-averse power-induced risk attitude on the part of the Roman Empire. The creativity of the classical Greek and Hellenistic eras soon disappeared during the reign of the Roman Empire. The extremely risk-averse power-induced risk attitude also caused severe distortions in the economic policy of the Roman Empire.

The commercial classes of the Roman Empire were small in size and enjoyed neither the wealth nor the status of the landowning aristocracy. Most production in the empire was small in scale, under-capitalized, and combined with a surprising lack of technological innovation. There was a reduction in trade, regionalization of economic activities, a growing number of deserted lands and a decline of urban areas.[94] In the late Roman Empire, the expensive navy was neglected and allowed to decline. Consequently, pirates plagued the Roman waters and trade suffered.[95] Although the Romans refined the area of law beyond what the Greeks had done, in other areas the empire suffered. The Roman culture, well known for its stress on discipline and seriousness of purpose, became a synonym of conspicuous consumption and public display. Due to massive economic mismanagement, the repression of commerce, and a decline in trade, the empire disintegrated into autarchic units. This trend of disintegration ultimately ended in political decentralization and dissolution of the western part of the empire.

The Roman Empire failed to conquer the whole Hellenistic world. Roman power was basically confined to the coastal regions of the Mediterranean Sea. The older seats of ancient civilizations, Mesopotamia and the Persian Highland, were largely out of the reach of the Roman legions. In the Middle East and Central Asia, which were once parts of the Hellenistic world, a revived Persia dominated the scene from 250 BC-226 AD as the Parthian Empire. The Parthian Empire was then succeeded by the even more powerful Sassanid Empire (226-651 AD).

The Parthian Empire had its origin in the northeastern part of modern Iran. The Parthian Empire defeated and deposed the long-declining Hellenistic Seleucid Empire, and reunited and ruled over the Iranian Plateau. At the height of its power, the Parthian Empire covered all of Iran proper, as well as regions of the modern countries of Armenia, Iraq, Georgia, eastern Turkey, eastern Syria, Turkmenistan, Afghanistan, Tajikistan, Pakistan, Lebanon, Israel, Palestine, Kuwait, the coastal regions of Persian Gulf of present day Saudi Arabia, Bahrain, Qatar, and the United Arab Emirates. The Parthian Empire was also in control of the Silk Road, the trade route between the West and China. The Parthian Empire was ruled by the Arsacid Dynasty, the third native dynasty of ancient Persia, after the Median and the Achaemenid dynasties. Practically uncontested within their geopolitical sphere, the Arsacid kings styled themselves "king of kings".

---

94 In contrast, the Eastern Roman Empire of the medieval era, in constant competition with the Sassanid Empire, undertook institutional reforms and survived much longer due to the pressure from Persia. Refer to Leppin (1998, pp. 249-262) in Bernholz et al. (1998).
95 Refer to Starr (1989, pp. 72, 117) on the economic decline of the Roman Empire.

The Parthians were nomads originally. In their long contests with the Seleucid heavy infantry combined arms legions, the Parthians had invented an effective and unique mode of warfare which became the prevalent way of war during the medieval era. Parthian military power was based on a form of the guerilla warfare of a mounted nomadic tribe, with light cavalry being the mainstay of the fighting force, and support from heavy cavalry and sizable light and heavy infantry. The Parthians were superb horsemen; the light cavalry were trained to shoot at full gallop while advancing or retreating. This was the famous "Parthian shot".

The political structure of the Parthian Empire was highly decentralized, partly due to the rather complicated terrain and the reliance on light cavalry. The earlier Seleucid Empire and the later Sassanid Empire were highly decentralized too: all three regimes were far short of the Achaemenid Empire in terms of political unity.[96] The constituent units of the Parthian Empire were highly independent, with many of the Hellenistic cities retaining much autonomy. That was the main weakness of the empire, causing it to lack the ability to mount sustained long-distance military operations: the independent regional lords whose support the Parthian monarchy depended upon were reluctant to travel far from their home power base for long periods.

The archrival of the Parthian Empire was the Roman Empire. The Parthian form of warfare suited the desert, oasis, steppe and highland topography of Middle East, Persia and Central Asia very well. In contrast, the Roman Empire relied mainly upon heavy infantry legions, supported by the navy as circumstances necessitated. The Roman Empire found it difficult to defeat the Parthian Empire, whose main battlefield fighting force was cavalry. Both the light and heavy Parthian cavalries were much faster and more mobile than the Roman foot soldiers.

In the Battle of Carhae (53 BC), Crassus, the Roman general who tried to conquer the Parthian Empire, was disastrously defeated by the Parthian commander Surena. This defeat testified to the limit of the Roman way of warfare in Middle Eastern terrain. The Battle of Carhae was a harbinger of the medieval form of warfare, when cavalry would become the dominant force on the battlefield.

The Romans had great trouble defeating the Parthians, but the reverse was also true. The Parthians found it difficult to conquer Roman eastern provinces, especially well-fortified areas guarded by the Roman heavy infantry legions. Furthermore, the great distance between the core regions of the two empires added to the difficulties of conquering and controlling the territories of the rival. A delicate balance of power was therefore imposed by nature. The two

---

96 Refer to Garthwaite (2005, pp. 66 – 68).

major powers, despite many clashes and turns of fortune, were largely confined to their own natural spheres of influence.

Though neither could truly threaten the survival of the other, some kind of political-military contest was however always going on between the Roman Empire and the Parthian Empire. The relations between the Parthian and Roman Empires consisted largely of intermittent warfare separated by periods of stalemated truce and exchanges of gifts and hostages. The fights were mainly over the borderlands between them, especially Armenia. In later interactions, Parthia was recognized by Rome as co-equal.

Several centuries later, the competition was more intense between the Sassanid Empire and the Roman (and later Byzantine) Empire. The Sassanid Empire (226-651 AD), the last pre-Islamic Persian Empire, defeated, succeeded, and became more powerful than the Parthian Empire: the traditional territory of the Sassanid Empire included all of present-day Iran, Iraq, and Armenia; the southern Caucasus (including southern Dagestan); southwestern Central Asia; western Afghanistan; parts of Turkey and Syria; coastal parts of the Arabian Peninsula and the Persian Gulf area; and some parts of southwestern Pakistan. For brief periods the Sassanid Empire also held Egypt, as well as Yemen and Oman, and controlled the maritime trade route with the east through the Red Sea, especially after conquering Yemen in 574 AD. Sassanid Persians established a military and political presence in present-day Yemen, for this choke-point between the Red Sea and the Indian Ocean could control the profitable trade between the Mediterranean and the Indian subcontinent.

From the third to seventh centuries AD, the Sassanid Empire with its heavy cavalry was the most important power in the Middle East, rivaled only by the Roman (and Byzantine) Empire. Compared with the Parthian military, heavy cavalry played a more important role in the Sassanid army. The mainstay of the Sassanid army was heavy cavalry, made up of elite aristocrats trained since youth for military service. The heavy cavalry was supported by light cavalry, infantry and archers. Administratively, the domains of the most powerful aristocrats functioned as semi-independent states.

Changes in relative power—the Roman Empire had weakened and become the Byzantine Empire, while the Sassanid Empire strengthened Persia—increased the competition between the two major cultures. The Romans and, later, the Byzantines treated the Sassanid Persians as the only equal power. Given the natural geography and the different forms of warfare that the Romans and the Persians specialized in, the mass factor between the two major empires was quite small. Consequently, the marginal effect of relative capability generated by the intermittent conflicts between the two empires was quite moderate and so was the development drive. The existence of a worthwhile rival and the small mass factor between the two major empires prevented the emergence of highly risk-averse power-induced risk attitude. Consequently, imperial complacency and conservatism had lesser chances to settle in both empires. The Persian culture had a renaissance during this time and the eastern half of the Roman Empire had its fair share of reforms—and of course eventually outlived the western half. The results were the Byzantine Renaissance and the Sassanid Golden Age of Persia.

Sassanid society, economy and culture were among the most flourishing of the late classical and early medieval era, matched in the Middle Eastern area only by those of the Byzantine civilization, their longstanding political and military rival. The level of scientific and intellectual exchanges between the two major cultures testifies to the competition and cooperation of these two major civilizations during their times. The Sassanid period saw the most advanced achievements of Persian civilization and had a great influence on medieval European and Asian (and especially Islamic) culture and art.

The Sassanid Empire had intensive development plans. Sassanid rulers consciously sought to revive Persian traditions and to eliminate Hellenistic cultural influence. Sassanid rule was known for its urban planning, agricultural development and technological improvements.[97] Canals were built and many cities were founded, some settled in part by migrants from the Roman territories, including Christians who could practice their religion freely under Sassanid rule. The Empire also had an amicable attitude towards Jews, who lived in relative freedom most of the time. (The official state religion was Zoroastrianism.)

This remarkable peak of Persian power was paralleled by a blossoming of Persian art, music, and architecture. Cultural expansion followed political influence and military victories, and Sassanid art penetrated Central Asia, reaching as far as China. The Academy of Gundishapur was a famous learning center with a great library. During this time the best pieces of ancient Persian literature were written and notable pieces of music composed, and sports such as polo became royal pastimes, a tradition that continues to this day in some regions. The time also saw the blooming of the indigenous religion of Zoroastrianism, which was a synthesis of several major religions and philosophies of the Middle Eastern, Western and Indian cultures.

Meanwhile, at the other end of the Eurasian continent, in China, the larger economies of scale in warfare due to the use of massive infantry legions and light cavalry set in motion a chain of changes that eliminated smaller states and created larger and larger states and alliances. Political and military competition was so intense that historians have called it the Era of the Warring States or the Era of the Contending States. After the Kingdom of Qin started a

---

97 Refer to Garthwaite (2005, pp. 99, 109).

series of reforms under the draconian legalist statesman Lord Shang (395 BC-388 BC), thus bolstering the war capacity of the state, the relative combined military and economic efficiency shifted significantly in favor of the Qin Kingdom. This asymmetry in relative efficiency and the large mass factor caused an irreversible tilt in the balance of power in favor of Chin. The ultimate outcome was an extremely high concentration of resources and capability in China; the unification of China under the First Emperor, Yin Zheng, King of Qin; and the establishment of the Qin Dynasty in 221 BC.

Though the Qin Dynasty was soon overthrown, the subsequent Western Han Dynasty swiftly reconstituted the imperial order. The late classical Chinese imperial order was highly stable and survived for over six centuries. It spanned four major dynasties: Qin (221 BC-206 BC), Western Han (206 BC-24 AD), Eastern Han (25-220 AD) and Western Jin (265-317 AD). The imperial order of late classical China was more complete than that of the Roman Empire: while the Roman Empire still had an equal in the form of Parthian and then Sassanid Persia, the Han Dynasty of China, however, practically had no equal.

The establishment of the imperial order in China, with its associated extraordinarily high concentration of resources and capability, reduced political-military competition, dampened relativist concern and sapped developmental energy. Given that the imperial regime nearly monopolized resources and capability, the larger mass factor greatly enhanced the power of the empire relative to possible challengers with much lesser resources and capability. Political-military competition was therefore further dampened. For instance, it was a popular belief that the Han Dynasty would last forever. Consequently, relativist concern took a further deep dip. Furthermore, the extraordinarily high concentration of capability in the hands of the imperial regime together with the large mass factor generated an extremely risk-averse power-induced risk attitude which bred strong conservatism and severely distorted economic decisions. The vigor, creativity and achievements of the pre-Qin competitive state system soon disappeared and were never repeated. In the Chinese lands, stagnation and decline resulted.

This period mainly saw the reign of the Han Dynasty. The core area of the Han Dynasty was the Northern Plain, centered along the middle and lower stretches of the Yellow River. This core area contained approximately two thirds of the total population, and its agriculture and economy formed the economic and military foundation of the imperial order. The Han Dynasty relied upon the massive infantry legions supported by this economy to hold the empire together and keep nomads and barbarians at bay, while a unified writing system facilitated the empire's bureaucratic administration.

Compared with the vigor of the preceding pluralistic Spring and Autumn Era and the Era of the Warring States, the Han dynasty was characterized by stagnation and decline. During Han China, the government deliberately repressed commerce. Economic growth was sacrificed for the sake of political stability.[98] The merchant class suffered a brutal loss in status and professional autonomy due to official policies and restrictions.

Except during the reign of Emperor Wu Ti, the ambitious sovereign who adopted a policy of actively pursuing and destroying the Xiongnu nomads who once harassed China, the Han Empire had a small and non-intrusive government that provided few public goods, in accordance with the teachings of Taoism in the earliest period and the teachings of Confucianism in later periods. A simple and self-sufficient agrarian society was the official ideal. Han Confucian scholars, for instance, severely criticized Lord Shang for the commercialization of land.[99] Government refrained from any intervention in the economy, including the provision of much-needed public intermediate inputs which, though not directly consumed, raised productivity and were critical for the functioning of the economy (but could not be effectively supplied by private initiative due to the problem of collective action or free-riding).[100] Consequently, productivity suffered, commerce declined and the economy deteriorated, especially the urban economy. Han Dynasty cities were generally smaller than their predecessors of the Warring States Era.[101]

Partly due to the faltering economy, the military capability of the Han Dynasty declined from that of the Era of the Warring States. In that earlier era one of the seven Warring States, the Kingdom of Zhao, effectively repulsed the encroachment of the nomadic tribal alliance of Xiongnu and conquered large tracts of territory in Inner Mongolia from the Xiongnu. Yet, it took the later and supposedly-stronger Han Dynasty Chinese Empire seven decades of preparation to successfully deal with the Xiongnu after suffering much humiliation, even though the Xiongnu threat consisted mainly of border incursions that did not truly threaten the survival of the Han dynasty.

On the cultural front, the Han Dynasty produced few great literary works of which to boast. Han Dynasty poetry was merely second-rate imitations of that produced during the Era of the Warring States. The policy of Emperor Wu to elevate the position of Confucianism to official orthodoxy suffocated both philosophical and political intellectual inquiries. In contrast to the First Emperor's bluntness in enshrining the orthodox teachings of Legalism by burning the

---

98 Refer to Wesson (1967, 1978), Huang (1988) and Hui (2005).
99 Refer to Raaflaub and Rosentein (1999, p. 39, footnote 83).
100 Refer to Stavrianos (1982, p. 148). Wesson (1967, p. 99) makes an economic contrast between the Era of the Warring States and the Han Dynasty.
101 This had been demonstrated by archaeologists. Refer to Raaflaub and Rosenstein (1999, p. 38, footnote 78).

books of the other schools of thought and burying alive the Confucian literati, the Han dynasty resorted to more subtle measures to elevate Confucianism to orthodoxy. Chinese minds of this era turned away from philosophic and intellectual exploration to mere superstition. Confucian orthodoxy also meant a small and non-interventionist government and, to ensure the smooth functioning of society and the polity, relied on moral teachings instead of law enforcement. Confucianism discouraged innovation and independent thinking as well as manual work or business dealings—and with such things discouraged, China's classical empires slowly declined.

4. South Asia

India was the exception during this era of universal empires. Geographically the South Asia subcontinent is composed of three major regions: the verdant Ganges River Valley, which is economically the most important; the mountainous south, which was the last to be transformed by civilization; and the arid Indus valley, which produced one of the earliest civilizations, the Harappan civilization. The great variations in South Asian's compartmentalized terrain have always posed onerous problems for empire building.

The first pan-South Asian Empire was the Mauryan Empire (321-185 BC). Originating from the Kingdom of Magadha in the Indo-Gangetic plains in the eastern side of the South Asian continent, the consolidation of the Mauryan Empire was partly due to the advantage gained when Chandragupta Maurya borrowed political know-how from the Macedonian invaders. The political know-how and philosophy of the Mauryan Empire was enshrined in the teaching of Arthashastra by Kautilya, and Chandragupta instituted a strong central administration. However, the effective rule of the Mauryan Empire lasted only for three generations, under Chandragupta Maurya the founder of the empire whose reign was from 322 BC to 298 BC, his son Bindusara who reigned from 297 BC to 272 BC and his grandson Ashoka the Great who ruled from 273 BC to 232 BC. In 185 BC, the last Mauryan emperor, Bridadrata, was assassinated by the commander-in-chief of his guard, Pusyamitra Sunga, who took over the throne and established the Sunga Empire.

The Mauryan Empire (322-185 BC) both began and ended earlier than the Qin and Han dynasties of China, the Parthian and Sassanid Empire of Persia and the Roman Empire. Around the time that the imperial orders began to establish themselves in China, the Mediterranean and Persia, the Mauryan Empire broke down into a state system functioning from around 200 BC to 300 AD. The major powers of this state system were the Kushan Empire in the Northwest, the Sunga Empire in the Northeast and the Satavahana or Andras Empire in the South.

These three major powers, especially the Sunga Empire and the Satavahana Empire, were states with compact and centralized structures.

The distribution of capability in South Asia during this era was highly even. Relativist concern was therefore high. Furthermore, the highly even distribution of capability prevented the rise of extremely risk-averse or risk-seeking power-induced risk attitudes to distort economic decisions. Consequently, there was in South Asia a significant intensity of interstate political and military competition and drive for development and creativity. The developmental drive generated by this state system propelled the Indian civilization to the pinnacle of its classical golden age.

Trade thrived in the Kushan Empire, whose territory extended from the Aral Sea through present day Uzbekistan, Afghanistan and Pakistan into northwestern India and controlled the link between the Silk Road and the maritime trade routes of the Indian Ocean. The dominance of the Kushan Empire over international trade was ended only when the rise of Sassanid Empire and Gupta Empire basically ended the rule of the Kushan Empire. The Satavahanas Empire greatly influenced Southeast Asia by spreading Hindu culture, language and religion into the region. The Sunga Empire had its share of achievements too. Within its realm saw the rise of the indigenous Mathura school of art, a counterpart to the Hellenistic influence in Indian art.

The arrival of the Gupta Empire ended the highly even distribution of capability within the Indian system. The Gupta Empire existed approximately from 320 to 550 AD and effectively governed northern India from 375-475 AD. The Gupta Empire was the last of the classical empires and the first of the medieval empires. From the fall of the Mauryan Empire (185 BC) to the establishment of the Gupta Empire (ca 320 AD), waves of nomads entered northwestern India. The Kushan Empire was in fact established by Indo-European nomads from central Asia. The establishment of the Kushan Empire and the invasions of nomadic tribes from Central Asia was the harbinger of the medieval era. Established at the end of an era dominated by massive infantry legions, the Gupta Empire relied more on cavalry, but the backbone of the Gupta military was still light infantry legions composed mainly of archers. The Gupta Empire was constantly under threat from the nomads from the northwest, especially the Hephthalites or White Huns.

The classical Indian state system did not come to an abrupt end when the Gupta Empire was established. The Gupta Empire was largely confined to northern India. More decentralized than the Mauryan Empire, it relied on matrimonial alliance with the Vakataka Empire of southern India for its influence in southern India and, to a large extent, its dominant position in northern India as well. Hence the Gupta Empire was a hegemonic power in a state system: the Gupta Empire's position was constantly contested and the system was not an unchallenged imperial order that monopolized military capability within an international political system. Given the constant threats that the Gupta Empire faced from the nomads from the northwest and the rival southern Indian states, it was less likely that imperial complacency and inertia could set in. The distribution of capability within the Indian system during the Gupta era was therefore moderately uneven or largely even. Given the largely even distribution of capability, the larger economies of scale associated with the use of massive light infantry legions in South Asia generated mostly positive scaling effect and very little negative unbalancing effect. Furthermore, since the increase in asymmetry in distribution of capability was quite moderate, the negative asymmetric effect was quite minor too. Consequently, the relativist concern during the Gupta era was at least as high as that of the state system of the previous era. The largely even distribution of capability also prevented the rise of extremely risk-seeking or risk-averse power-induced risk attitudes to distort economic decisions. This pluralistic international order of India with its constant political-military competition in turn enabled great developmental achievements to continue.[102]

There were great advances in science and arts. The Gupta Empire was a great patron of education and supported the Universities of Nalanda and Vikramasila. During this era, Sanskrit learning was revived to serve as a lingua franca and Sanskrit literature flourished.[103] The two great epic poems of India, *The Mahabharata* and *The Ramayana*, achieved their final form during this era. There was the compilation of the authoritative Hindu Law Books (Manava Dharmashastra), the foremost of them being the Code of Manu. In the second century, Charaka compiled the oldest surviving Indian medical textbook, the Charaka Samhita; the great linguist, Panini, brilliantly analyzed the physiology and morphology of the Sanskrit language in his "Eight Chapters" (Ashtathdyayi), and then Patanjali's "Great Commentary" (Mahabhashya) on the "Eight Chapters" was completed. Sanskrit was therefore the first language to be scientifically analyzed. The decimal place system of numerical notation, one of the great inventions of the human mind, emerged in India by 270 AD. In art, after centuries of Hellenistic dominance, an indigenous Indian style flourished. Politically, economically and culturally during this period southern India too was on the rise. One of the southern India power, the Chola Empire, was a thriving trading state, with its navy being the mightiest in the Indian Ocean. It conquered the Maldives and defeated the fleet of Srivijaya, the great maritime empire of Southeast Asia that spanned present-day Malaysia and Indonesia.[104] This period saw the Indian colonization of Southeast Asia, while in Southeast Asia, beginning

---

102 Refer to Stavrianos (1982, pp. 134-137).
103 Refer Kulke and Rothermund (1998, pp. 79-81).
104 Refer to Wolpert (2004, pp. 68-113).

at about the dawn of the Christian era, there was a massive acceptance of Indian culture.[105] Local states in Burma, Sumatra, Java, Malaya, Siam and Vietnam eagerly imported as much Indian civilization as they could. Hindu states and empires established in Southeast Asia included the Majapahit of Java and the Khmer of Indochina.

International trade prospered with extensive trade to the west, to Southeast Asia, to China and to Central Asia. Indian cultural influence spread far abroad, especially to Southeast Asia, and a Greater India emerged. Buddhism was established in Southeast Asia, Central Asia and East Asia. In Northeast Asia, Buddhism, with its accompanying art forms, flowed like a mighty torrent into China, Korea and Japan in the period ca 200-600 AD.[106] The period from ca 184 BC-320 AD and the subsequent Gupta Era were times of political fragmentation in India, but therefore also times of economic enrichment and cultural splendor.

5. Eurasia

During the late classical era an inchoate mega-Eurasian international political and economic system was taking shape. Major empires or great powers from different civilizations started to interact culturally, economically, politically and militarily. How did such interactions of a pan-Eurasian scale affect the relativist concern and power-induced risk attitudes of the major empires or great powers and their developmental performance?

The political and military rivalry between Persia and the Roman Empire was the most important among such interactions. It was also unique, for the interactions between other pairs of major empires or great powers were either non-political or non-military in nature, or were much less frequent and on a much lower level of intensity. For instance, though China and Persia had constant contacts during late classical times, there was however no political-military competition between them. The great distance between the core regions of the two empires meant that both faced strong diminishing returns in establishing political and military influence in the geopolitical niche of the other. That is, there was a very small mass factor or low levels of economies of scale in conflict between the two major cultures. Therefore, trade and cultural exchanges were the main concerns of that relationship. That being the case, the relativist concern between

---

105 Refer to McNeill (1999, p. 274).
106 Refer to McNeill (1999, pp. 180-193).

the Parthian Empire and the Chinese Han Dynasty of China was very low or non-existent. Similarly, though the Romans and the Chinese were aware of the existence of each other, there was no political and military contest between them due to the great distance between the two cultures.

The Mauryan Empire had a conflict with the Seleucid Empire in 305 BC. However, given the vast distance between the core areas of the two empires, the Mauryan core area was in the Eastern part of north India and the Seleucid Empire had its core in the Mesopotamia and Anatolia, it was difficult for either to conquer the other or to have a long protracted war. Consequently, Seleucus and Chandragupta sealed a treaty in 305 BC through which the Seleucus ceded a number of territories to Chandragupta, including southern Afghanistan and parts of Persia. In return, Seleucus obtained five hundred war elephants, a military asset which proved useful in his latter campaigns.

The interactions between the Persian Parthian Empire and the Kushan Empire of Central Asia and India were of a similar nature centuries later. The Parthian Empire made significant inroads into India, controlling extensive territory in modern day Pakistan after defeating the Kushan Empire. However, given the great distance from the Parthian core area and the decentralized administrative structure of the Parthian Empire, this far-flung province soon declared independence and became the Indo-Parthian Kingdom. Since there was such a great distance between the core areas of Persia and India and, there was also a significant difference in the modes of warfare suitable for the two regions, the mass factor between Persia and the Indian states was quite small. The interactions between Persia and India were therefore somewhat similar to the contests between Persia and Rome, though at a lower level of intensity. Consequently, the sporadic interactions between the Parthian Empire and the constituent members of the Indian state system were not significant enough to generate high relativist concern or to spur great development effort.

Summing up, the great distance among the major cultures made wars of conquest between them an impractical option during the late classical era. Consequently, the distribution of resources and capability among them was determined more by natural endowment than by political and military contests. This geopolitical compartmentalization of the major cultures would be broken up only in the medieval era due to progresses in military technology, especially the rise of heavy cavalry which created greater projection of power for the military.

6. Conclusions

Summing up, the heavy infantry military technological revolution led to structural changes in international politics of the late classical major Eurasian civilizations.

India, with its more even distribution of capability and higher intensity of political military competition, was more creative and had better developmental performance. The heavy infantry military revolution facilitated the rise of the late classical era gigantic empires in the Mediterranean world, Middle East and China. Consequently, imperial complacency and conservatism reigned in these diverse lands and sapped the energy of progress.

CHAPTER 6
Abbasid Golden Age and Sung Puzzle

1. Introduction

This chapter analyzes the long medieval era, from approximately 400 AD to 1300 AD. The defining military technological change of this era was the rise of the heavy cavalry which led to a small mass factor. The rise of heavy cavalry also led to the advance of nomads at the expense of settled societies, and the universal empires of the late classical era either retreated or dissolved. Fragmentation and creativity were the two prevalent themes among the major medieval cultures of Western Europe, the Byzantine Empire, South Asia, China, Japan and the Islamic world. Given the small mass factor, relativist concern was generally quite modest in the major cultures despite the rather even distribution of capability. The medieval major cultures therefore functioned significantly different from the familiar early modern European competitive state system with its strong emphasis on real politics and high relativist concern.

The medieval Islamic civilization deserves special attention here. The cultural energy and economic achievements of the Islamic civilization during its early era of unity and then the Abbasid Golden Age contradict the Hume-Kant hypothesis of Bernholz et al. (2004). According to the Hume-Kant hypothesis, one would look at the medieval Islamic world and expect to see a state system with independent states of roughly equal power. Similarly, the above average performance of China during the mighty Tang Dynasty also poses problems to the Hume-Kant hypothesis. The Tang Dynasty was definitely not a state system and was in fact about as powerful as the Han Dynasty before, or the Ming and Qing Dynasties that came later. By taking into account the effects of military technology on international political structure and, political military competition and economic performance, the arguments developed in earlier chapters successfully accounts for the achievements of the medieval Islamic world and the above average performance of Tang Dynasty China. The Chapter also treats the Song Puzzle in greater details. The high economic achievements of Song Dynasty China resonates very well with the early modern European Miracle and strongly supports the claims of Jones (1981, 1988, 1990) that intensive growth was a recurring phenomenon in history.

The medieval era saw more significant political military competition at the pan Eurasian level. Besides the struggle between the Christendom and Islam in the Middle Eastern and the Mediterranean regions which in a way was a successor to the clash between the Romans and the Persians, there were many more political military interactions among the major civilizations. During the medieval era, the Abbasid Caliphate clashed with Tang China over central Asia, Islamic forces began their conquest of South Asia, Tang China and Japan fought in Korea, Tang China briefly exerted military influence in India, the southern Indian state of Chola sent naval expeditions to Southeast Asia, among others. These political military interactions at the pan Eurasian level culminated in the far reaching conquests of the Mongol Empire, the first truly pan Eurasian empire in history.

2. Heavy Cavalry

The Eurasian steppes saw the advent of selective horse breeding, which led to a powerful steed able to carry a fully armed and armored warrior at great speed, even when the horse itself was fully armored—and these newly-bred horses could then charge with great explosive power. The invention and use of stirrups and composite bows added further power to the cavalry: horse and rider were effectively welded into a lethal fighting unit. Consequently, the earlier battlefield reign of the massive light and heavy infantry legions came to an end, and until the gunpowder military revolution, the mounted shock combat of the heavy cavalry in combination with the mobile archers on horseback of the light cavalry ruled the battlefield. The Persians, in their long struggles against the Seleucid phalanxes and the Roman legions, pioneered this wave of military technological revolution, and soon the practice spread to the other parts of the Eurasian landmass. Naturally, peoples already used to riding on horses particularly embraced this military advance.

As the massive infantry legions which the universal empires of the classical agrarian world had relied upon lost their advantage in battle against the nomads on horseback, the steppe peoples gained greater military efficiency relative to the settled societies. Consequently, nomadic hordes advanced at the expense of the settled societies. The rise of the cavalry as the dominant military force changed not only the relative combined military economic efficiency between settled societies and nomads. This wave of military technological changes also reduced the returns to scale in warfare since cavalry relied less upon manpower and military mass in battle. Since the ancient agrarian empires had much more resources at their disposal than the steppe peoples, a decline in the mass factor further reduced the military advantage of the once-mighty agrarian classical universal empires, which therefore either retreated or dissolved. This wave of military technological changes affected all major Eurasian civilizations.

Nomadic invasions led to either the collapse or the retreat of the classical universal empires around 400-600 AD. The aftermath was political fragmentation and decentralization. The decline of the Roman Empire, the Sassanid Empire of Persia, the Gupta Empire of India, and the Chinese Western Jin Dynasty happened almost simultaneously.[107] The invasions came from many sources: Asiatic nomads from the steppe of central Asia, Arabian tribes from the desert of the Arabian Peninsula, and Germanic tribes from the northern and central European forests and plains. Massive empires collapsed into small kingdoms which in turn dissolved into tiny principalities.

The ascendancy of the cavalry over the infantry and the associated decrease in the economies of scale in conflict had economic and public financial consequences. Cavalry warfare was horse-intensive and had less need of manpower; pasture was the critical resource that supported the military on horseback. The fiscal capacity of the state became less important. Centralized states gave way to feudalism. Localized feudal relationships and exchanges replaced centralized unitary taxations and economic regulations and networks. This decentralization in the economic and fiscal arenas led to further fragmentation of the medieval geopolitical landscape. The consequence was a myriad of small kingdoms or short-lived and decentralized empires that came and went as charismatic or extraordinary leaders entered and exited the historical stage.

The heavy cavalry military revolution indirectly affected the relative combined military and economic efficiency of the major Eurasian cultures through another mechanism. Since the medieval form of warfare relied less on manpower but critically upon horsepower, the control of a dominant economic core area with abundant manpower became less important for maintaining military might. Consequently, even in regions where there was a single dominant core economic area with a concentration of an overwhelming share of resources and manpower, like China, the relative combined economic and military efficiency between the contestant controlling the dominant core economic area and contestants holding other lands became less asymmetric. The player controlling the dominant core now enjoyed less strategic advantage, since economic resources and a greater manpower supply could translate less into military might. As the relative combined economic and military function became more symmetric, there was a more even distribution of resources and capability within all the major cultures of the medieval era.

In order to meet the threat from the steppe nomads of Central Asia, Parthian and Sassanid Persia had replaced infantry legions with localized heavy and light cavalry as the core of its defense. Persia therefore decentralized its public

---

107 Refer to McNeill (1982), Dudley (1990, 1991, 1992) and Keegan (1993).

administrative apparatus. The Byzantine Empire partly emulated the Persian defense system: the decrease in the economies of scale in warfare led to decentralization and retrenchment of the Byzantium government in the 7th century AD, which in turn led to the rise of feudalism.[108] However, even the improved defense system could not protect Sassanid Persia against the onslaught from another group of nomads, the Islamic Arabs.[109] In 632 AD, the first raiders from the Arabian tribes, recently united by Islam, reached Sassanid territory. Years of warfare had already exhausted both the Byzantines and the Sassanids. The Sassanids were further weakened by the increasing power of the provincial feudal lords and a rapid turnover of rulers. These factors facilitated the swift Islamic conquest of Persia.

The core area of the Roman Empire was the Mediterranean basin. The Roman Empire held together the many coastal regions of the Mediterranean Sea with its great fleets and massive infantry legions. With the rise of cavalry and the decline of infantry in warfare, it became nearly impossible to hold both the European northwestern coastal regions and the Asian-African southeastern coastal regions of the Mediterranean together under one empire. The Roman system was divided into two: the Western Roman Empire, and the Eastern Roman Empire, otherwise known as the Byzantine Empire. This division was soon followed by dissolution as the Western Roman Empire subsequently collapsed under waves of nomadic assaults, whilst the Byzantine Empire, despite the heroic attempts at resurgence by the Emperor Justinian, retreated under incessant nomadic invasions.

Under repetitive onslaughts from the nomadic tribes, the major classical empires of the settled civilizations abandoned the plains and retreated to mountainous and/or maritime regions of their former realms. These included the Eastern Jin Dynasty, the Southern Dynasties and the Southern Song Dynasty of China; the Hindu southern Indian states that confronted the Muslim powers which invaded and conquered the Indo-Gangetic Plain; the Byzantium Empire; and the Italian city-states that controlled much of the Mediterranean trade. It was the Deccan Highland in India and the Yangtze River in China that served as natural barriers to the power of the nomadic cavalry. As a response to the supremacy of cavalry warfare, the Byzantine and Persian armies increased the use of cavalry and came to look very much the same as the nomadic hordes. These remnant successor states of the once mighty late classical empires of the settled civilizations managed to long endure in regions where the military superiority of the cavalry was limited or annulled by geography.

---

108 Refer to Treadgold (2001, Ch. 4).
109 Refer to McNeill (1999, pp.194-209).

The medieval era refers to the period from the collapse of the classical universal empires to the gunpowder military revolution that kick-started the modern era, that is, from around 400-1300 AD. Given the low level of economies of scale in conflict, fragmented pluralistic international orders—interrupted briefly by short-lived and decentralized empires—were the norm for the geopolitical landscape of the major Eurasian cultures. Stable, lasting and centralized imperial orders did not reappear on the Eurasian geopolitical scene till the arrival of the gunpowder military revolution.

During the long medieval era, there were many brief ambitious imperial attempts across Eurasia. In Europe, the short reign of limited unification under the Charlemagne Empire lasted 768-ca 850 AD. In the Middle East, the newborn Islamic Empire was united under the Umayyad Caliphate for a brief period of about a hundred years, from around 661-750 AD. In India, after the collapse of the Gupta Empire in the fifth and sixth centuries, political unification did not return until the arrival of the Delhi Sultanate, which briefly unified northern India in the thirteenth century. In China, the Tang Dynasty brought about a relatively short-lived unification from 618-755 AD. These brief imperial attempts pale in comparison to the long lasting uncontestable and stable imperial orders of the preceding late classical eras or the following gunpowder era.

The small mass factor of the medieval era had great political and geopolitical impact. For instance, Tibet, a state with a very small population and limited resources, sacked the capital of the Tang Dynasty three times. The Tibetan Empire ruled part of northeastern India as well but collapsed in around AD 850.[110] Empires of this era had short reigns and were more contestable and unstable, and the international political orders were thereby less likely to suffocate creativity and developmental momentum. The unstable, contestable and short-lived empires of the medieval age were unlike the stable, uncontestable and enduring imperial orders of the ancient world. Nor were they similar to the stable, uncontestable and very secure continental-sized empires of the gunpowder era. These medieval empires, given their insecurity and decentralized structure, were never as entrenched and complacent as the classical empires or the gunpowder empires. Furthermore, their reigns were not long enough to choke off creativity and developmental momentum, which was in fact generated by the interstate rivalry of the pluralistic and fluid international orders which ruled the Eurasian world for most of the medieval era. A time of international instability and fragmentation, the medieval era was also, somewhat surprisingly, a time of innovations, development and progress.

---

110 Refer to Grousset (1970, pp. 80-114) and McNeill (1999, pp. 224-227).

The creativity and energy of the medieval civilizations were not as splendid or spectacular as those of the classical Greek, Ganges Indian and Chinese civilizations or the early modern European civilization. For given the low economies of scale in warfare, political military competitions within the different medieval major cultures were also less intense compared to those of the classical Greece, Ganges India and China or early modern Europe. Nonetheless, all major medieval cultures developed and there were no spectacular differences.[111] Low economies of scale in warfare caused all major cultures to operate under pluralistic international orders, such as state systems, during most of the medieval era. Consequently, developmental performance was much the same across Eurasia. McNeill (1963) termed it "the rough equilibrium" among the major cultures.

One manifestation of the medieval energy and creativity was the expansion of the various civilizations. Christendom expanded to Northern and Eastern Europe.[112] Islam expanded into sub-Saharan Africa (after 1000 AD), eastern Africa, India, Central Asia, and Europe; after 1200 AD, Islam spread to Southeast Asia as well.[113] China exerted its influence into Yunnan, Manchuria, southern Siberia, Central Asia, Taiwan, and the Ryukyu Islands, and spread its cultural influence into Japan. The southern India Hindu states expanded into Southeast Asia.[114] Each major culture will be surveyed in turn.

3. Middle East

The conquests of the nomads brought a new major player into the Eurasian world of civilizations: the Islamic civilization. Bursting out from the Arabian Desert with great religious zeal and unity, the Arabian tribesmen under the banner of Islam soon conquered Sassanid Persia and wrested from Byzantium the wealthy provinces of Egypt, Syria and the Levant. From 661-750 AD, the Umayyad Caliphate conquered and unified the Iberian Peninsula, northern Africa, the Middle East and Central Asia under one faith. The warmer and more arid southern and eastern coastal regions of the Mediterranean world were now separated from the colder and more forested Western world, a separation that persists to this day. Yet the unified Islamic empire did not last long.

---

111 Refer to Abu-Lughod (1989) and Chaudhuri (1990).
112 Refer to McNeill (1999, Ch. 14, 16).
113 McNeill (1999) observes "In East Africa, the effort required to resist Islamic assault actually generated something of a 'golden age' in Abyssinia's cultural history......" (?77)
114 Refer to Haywood (1997, pp. 32-33).

The Arabs were originally nomads and therefore initially enjoyed high relative combined military and economic efficiency versus their settled rivals given the medieval form of warfare in which the cavalry played a dominant role. This high asymmetry in relative combined military and economic efficiency facilitated the creation of an empire. Furthermore, the initial strong appeal of a new universal religion and the strong bond it created among the Arabian tribesmen generated a short-lived condition of even higher asymmetry in relative combined military and economic efficiency. These favorable conditions generated a very high concentration of resources and capability and made possible the century-long unification of the Islamic world under the Umayyad Caliphate. However, political unity was rare during the medieval era, due to the small economies of scale in warfare of the medieval military technology. As the initial religious zeal faded and the Arabian tribesmen settled and lost some of their equestrian skill, relative combined military and economic efficiency became more symmetric. Consequently, the distribution of resources and capability also became more even and the unified Islamic empire began to breakdown.

With medieval military technology it was difficult to maintain a unified imperial rule throughout the Islamic world (or indeed anywhere across Eurasia) for long. After the Umayyad Caliphate, political fragmentation was the norm in the Islamic world. During the long rule of the succeeding Abbasid Caliphate which lasted from 750 AD to 1258 AD, political fragmentation was in fact the norm in the Islamic world. The Islamic world under the Abbasid Caliphate could be aptly described as a state system or a pluralistic international order, for among the many contestants in the Middle Eastern Islamic world, there was a fairly even distribution of capability. It was during this period of political fragmentation that saw the realization of the great Islamic achievements.[115]

The Abbasid Caliphate failed to unite the whole Islamic world: Egypt was autonomous, as were the Aghlabids of Tunisia and Algeria, and the Persian dynasties of the Tahirids and Saffarids. Furthermore, a descendent of the Umyaad Caliphate escaped to the far flung Spain to reestablish the Umyaad rule there. Above all, the Abbasid Caliphate still had to contend with the Byzantine Empire on the international scene. The reliance on Turkish slave soldiers for defense and regime maintenance also meant that the Abbasid Caliphate was far from monopolizing the capability of coercion and conquest. The authority of the Abbasid Caliphate was more religious and de jure than political and de facto. Consequently, there was a pluralistic international political order in the early

---

115 Refer to Ruthven (2004) for the geopolitical landscape of the Islamic world in history. Refer to Huff (2004) in Bernholz and Vaubel ed. (2004, pp. 204-210) for the Islamic golden age of scientific creativity during this era.

and mid-medieval era in the eastern Mediterranean and Middle East-Persia region.[116]

The following centuries saw a number of states came and went in the Middle East. In the mid- and late medieval era came the Turkish empires of the Ghaznavids and Seljuks. In 909 AD the Fatimid Dynasty was established in Tunisia and Egypt. Not much later the Ayyubid Dynasty was installed in Egypt and Syria, followed by the Egyptian Mamluk Dynasty, the Buyid Dynasty (established 945 AD) in Iraq, the Qarmatians in Bahrain, the Samanids in Eastern Persia, the Hamdanids in Northern Iraq, and many states set up by the Seljuk Turks. The fluid geopolitical landscape was frequently and swiftly reshaped again and again, mainly due to the small economies of scale in medieval warfare which made political regimes highly contestable. Consequently, complacency had very little chance of sinking in.

The enervation of the Arabian tribesmen created a vacuum in the military sector of the Islamic empire which was swiftly filled in by soldiers of Turkish origin. Although individual Turkish generals had already gained considerable, and at times decisive, power in Mesopotamia and Egypt during the tenth and eleventh centuries, the coming of the Seljuks signaled the first large-scale penetration of Turkish elements into the Middle East. Descended from a tribal chief named Seljuk, whose homeland was beyond the Oxus River near the Aral Sea, the Seljuks developed a highly effective fighting force. Furthermore, through their close contacts with Persian court life in Khorasan and Transoxania, the Seljuks also attracted a body of able administrators. Extending from Central Asia to the Byzantine marches in Asia Minor, the Seljuk state under its first three sultans—Tughril Beg, Alp-Arslan, and Malikshah—established a highly cohesive, well-administered Sunni state under the nominal authority of the Abbasid caliphs at Baghdad. This and later Turkish states coexisted with the many other small Islamic kingdoms in the Middle East and Northern Africa during the long medieval era.

The Islamic pluralistic international order was maintained until the Mongol invasion of the Middle East at the dawn of the gunpowder military revolution. Therefore, there was an increase in the symmetry of relative capability in the transition from the moderate or highly stable imperial order (as seen under the earlier Parthian and Sassanid Empires) to the medieval Islamic pluralistic international order. The medieval era was a period of frequent political chances and contests and wars for the Islamic world, but it was also a period of dynamism, energy and achievements which culminated in the Abbasid Golden Age.[117]

---

116 Refer to Goldschmidt (2002, Ch. 6).
117 Refer to Goldschmidt (2002, Ch. 7).

Politically, culturally, and economically, there were many achievements under the Abbasid Caliphate. This period experienced the Arab agricultural revolution, with the widespread diffusion of new crops and the promotion of new or the rehabilitation of old irrigation systems.[118] Trade was thriving and Arabian traders were at the center of the global trading network. Arabian and Islamic trading communities could be found throughout Eurasia: in Europe, Central Asia, India, Southeast Asia and China. Islamic merchants pioneered many innovations in business organization and finance: capital was pooled through partnerships, and letters of credit and promissory notes facilitated long distance trade.[119] There was intensive economic growth and a high level of urbanization and trade.[120] The city-states of Islamic Spain for instance achieved splendid cultural achievements and a high level of economic prosperity during this period.[121]

Politically, this period saw many outstanding statesmen and administrators. For example, the Persian Nizam-al-Mulk was one of the greatest statesmen of medieval Islam. For twenty years, especially during the rule of Sultan Malikshah, he was the true custodian of the Seljuk state. Besides being a great administrator, he was an accomplished writer and well known for his style: his book on statecraft, *Siyasat-Namah*, stresses the responsibilities of the ruler and is a valuable source for the political thought of the time. For instance, he argues that if a man is killed because a bridge is in disrepair, it is the fault of the ruler: he argues that the ruler should make it his business to appraise himself of the smallest negligence of his underlings.

Nizam-al-Mulk was a devout and orthodox Muslim who established a system of madrasahs, or theological seminaries (also called nizamiyah, after the first element of his name). These madrasahs provided students with free education in the religious sciences of Islam, as well as in the most advanced scientific and philosophical thought of the time. Many great scholars taught in these seminaries. For example, the famous theologian al-Ghazali (whose greatest work, *The Revival of the Sciences of Religion* was a triumph of Sunni theology) taught for a time at the nizamiyah schools in Baghdad and in Nishapur. Furthermore, Nizam-al-Mulk was the patron of the poet and astronomer Umar al-Khayyam (Omar Khayyam) whose verses, as translated by Edward FitzGerald in the nineteenth century, have become almost as familiar to English readers as Shakespearean sonnets.

---

118 Refer to Watson (1974).
119 Refer to Cahen (1970, pp. 511-38) in Holt, Lambton and Lewis ed. (1970) and Abu-Lughod (1989).
120 Refer to Jones (1988).
121 Refer to Wesson (1978, pp. 93-95), Abu-Lughod (1989), Armstrong (2000, pp. 45-81), Sonn (2004, Ch.2) and Bernholz (2004, pp. 184 - 193) in Bernholz and Vaubel (2004).

On the legal front, the Abbasid Era was the period in which Islamic law developed on a broad basis and the overall framework was established for all aspects of legal institutions. Laws relating to commerce and property were especially highly developed. The Islamic world during this period had already developed advanced legal concepts and practices on bankruptcy, partnership, trust, conveyance, and inheritance. That greatly facilitated the functioning of the highly commercialized economy. However, the overall development of Islamic law ended around the 13th century, a time that roughly coincided with the ending of the medieval Abbasid Golden Age of the Islamic world. Stagnation in legal thinking then dominated the next few centuries in the Middle Eastern world.[122]

Scientific achievements of the Abbasid Caliphate were noteworthy. In Baghdad, there was caliph al-Ma'mun's Bayt al-Hikmah, a great learning center. Astronomy and mathematics saw great advances: the period saw the manufacturing of astrolabes and the building of planetariums, and Islamic scientists knew that the earth orbited the sun earlier than Copernicus or Galileo did. The greatest Islamic mathematician, Muhammad ibn Musa (780-850 AD), wrote on the Hindu numeral system and popularized it—and as a result the system we use today is known, somewhat unfairly, as the Arabic numeral system. He compiled a textbook on algebra (itself a word derived from Arabic), and this book was used in both East and West for centuries; he also formulated the oldest trigonometric tables known, and collaborated with other scholars in preparing an encyclopedia of geography.

Advances in medicine and chemistry were equally splendid. In medicine there was the well-known Vesalius.[123] Ibn-Sina (Avicenna) wrote a widely used *Canon* for medicine.[124] Sophisticated laboratory techniques for handling drugs, salts and precious metals were developed.[125] Of all fields of medicine, Islamic achievements were greatest in ophthalmology. Physicians performed sophisticated surgeries including the earliest removal of cataracts. In the field of history, Ibn Khaldun, the North African polymath, produced his great work *Kitab al-Ibar* on universal history of which the famous first volume, *Muqaddimah*, is known in the West as *Prolegomenon*.

Collectively, the economic, political, cultural, legal, and scientific achievements of the medieval Islamic civilization were so great that they profoundly affected the course of human civilization and overshadowed those in the later periods of Islamic history. This pluralistic era of the Islamic world, known for its great cultural and economic achievements, is therefore known as the Abbasid

---

122 Refer to Wigmore (1928, pp. 541-542, 555-573).
123 Refer to Goldschmidt (2002, Ch. 8).
124 Refer to McNeill (1999, p. 246).
125 Refer to Stavrianos (1982, pp. 23-235) and McNeill (1999, pp. 219-220).

Golden Age. Through conquests, trade and cultural exchanges, Islamic influences during this period expanded from Middle East into South Asia, Central Asia, Southeast Asia, Europe, Sub-Saharan Africa and East Asia.

4. South Asia

India was affected by the heavy cavalry revolution quite early. The Gupta Empire was weakened by nomadic invasions soon after establishing its hegemony, and the South Asian subcontinent sank into prolonged political fragmentation following the retreat and collapse of the Gupta Empire. Hordes of nomads from Central Asia invaded north India in the early 6th century AD.[126] The succeeding Harsha Empire ruled over only parts of northern India, and as a whole, the subcontinent operated under a pluralistic international political system until the brief unification under the Delhi Sultanate (ca 1200 AD).

Compare to medieval India, late classical India including the Gupta era also had a largely even distribution of capability. The decrease in economies of scale in warfare following the rise of the heavy cavalry therefore reduced political military competition in India. Given the multiple core areas of India, interregional warfare was mostly aimed at the control of intermediate regions, or was simply for the acquisition of goods or prestige. A rough balance of power was maintained: rulers were unable to extend their control beyond their respective regions as their military equipment, administrative machinery and strategic concepts were roughly equivalent. However, given the rather equitable distribution of military capability, there was still considerable concern for overall state power. Furthermore, the establishment of Islamic states in northern India increased pressure on Hindu states in southern India. Under such competition, there was still considerable developmental drive and effort taking place, albeit not at the level of the earlier Gupta golden age era.

On the whole, developmental achievements did not lag far behind those of the Arabian or Chinese systems. The south Indian Chola Empire and Vijayanagar Empire, for instance, were renowned for their high level of prosperity brought about, and maintained by, extensive external maritime trade and a powerful army and navy. The Vijayanagar Empire, 1336-1565 AD, had a powerful navy and dispatched naval expeditions to Southeast Asia. Its predecessor, the maritime south Indian Chola Empire, was also a thriving trading empire that had sent naval expeditions to Southeast Asia. There was a strong Indian influence

---

126 Refer to Kulke and Rothermund (1998, pp. 89-91).

in Southeast Asia during this era.[127] The southern Indian power of the Cholas conquered Sri Lanka and sent both troops and trading fleets to the Ganges region, Indonesia, Malaya and the Maldives, enabling the great merchants of south India to play a central role in international trade.[128] Culturally, this period produced Bana, one of the greatest Sanskrit writers.[129]

5. The Byzantine Empire

The medieval ascendancy of heavy cavalry reduced the relative combined military and economic efficiency of the Byzantine Empire versus its many enemies, especially the nomadic tribes and Middle Eastern powers with strong cavalry. The associated decrease in economies of scale in warfare further weakened the Byzantine strategic position. The wealth and resources of the Byzantine Empire mattered less militarily. The Justinian effort to maintain the glory and integrity of the Roman Empire under a vastly different and much more disadvantaged military technology ultimately failed. Western Roman territories recaptured under Justinian rule failed to be held for long, and most were in fact lost again before the end of Justinian's reign. Consequently, the military technological changes resulted in a more even distribution of resources and capability in the Byzantine geopolitical sphere during the medieval era.

The Byzantine Empire lost its command of the preponderant share of resources and capability within its geopolitical sphere. Greatly weakened and shrunken, the Byzantine Empire had a higher relativist concern than the mighty Roman Empire had had. In the late classical era, the Roman Empire enjoyed an extremely high asymmetrical distribution of capability and a larger mass factor: it had become a very secure, stable and uncontestable empire with a very low relativist concern and an extremely risk-averse power-induced risk attitude. The change in fortune from the Roman Empire of the late classical era to the medieval Byzantine Empire therefore caused an increase in relativist concern and a more moderate risk attitude. There were therefore greater efforts in development and fewer distortions in economic decisions. So while the deteriorated political military condition may initially seem negative for Byzantium, the result was in fact the so-called Byzantine renaissance—a rebirth that left behind Roman imperial conservatism, indolence and insolence, Roman vices that Emperor Justinian deplored and campaigned against.

---

127 Refer to Haywood (1997, pp. 30-33).
128 Refer to Kulke and Rothermund (1998, pp. 109, 115-151).
129 Refer to Kulke and Rothermund (1998, pp. 103-107).

The Byzantine economy was among the most prosperous and advanced in Europe and the Mediterranean world. The most important economic pillar of the empire was trade. Constantinople was the center of a trading network that spanned practically the whole of Eurasia and North Africa. The legislative work completed during the reign of Emperor Justinian, *Corpus Juris Civilis*, an extensive revision of the ancient Roman legal code remarkable for its sweeping character, is a collection of laws that came to be referred to as "Justinian's Code" and was instrumental in inspiring the revival of Roman laws in Western Europe during the high middle ages.

The Byzantine renaissance was not an exception during the medieval era. Under incessant assaults from the nomadic tribes, often viewed as barbarians by the settled societies, the deteriorating military environment prompted increased efforts on the part of the retreating classical empires, or their remnant successor states, to compensate for their declining strategic superiority. Around 600-700 AD, both the Byzantine Empire and Sassanid Persia increased their developmental efforts and generated some kind of renaissance and resurgence. Good examples include the formation of Alexandria's New Aristotelian Academy and Sassanid Persia's Jundishapur as famous learning centers.[130]

The once-great states, now under threat from "barbarian" cavalry, strove to acquire security through greater economic might and indeed, some managed to hang on even until the arrival of the gunpowder era. The Byzantines experienced a strong renaissance during the mid-medieval era partly due to political competition with the Islamic world.[131] Led by Greek Christians in the Byzantine Empire, this vigorous Hellenic renaissance was ended by the Ottoman Turks in 1453 AD, but echoes of the Hellenic renaissance continued to be felt in Italy, whose own renaissance was significantly stimulated by the earlier Byzantine resurgence.[132]

6. Europe

With the rise of Islam, Christendom lost the North African coastal regions of the Mediterranean basin, together with the Levant and the bulk of the Iberian Peninsula. Of the vast realm of the Roman Empire, medieval Christendom retained basically only France and Italy, besides those lands still held by the much reduced and humbled Byzantine Empire. Christianity, however, compensated by adding northwestern Europe to its domain.

---

130 Refer to Goldschmidt (2002, Ch.8, p. 136).
131 Refer to Huff (1998, pp. 204-210) in Benholz and Vaubel (2004).
132 Refer to McNeill (1999, pp. 250-251).

With North Africa and the Middle East lost and northwestern Europe added, Christendom, unlike the Roman Empire, had no single dominant core area. Christian Europe was geographically more fragmented than China, India, and the Middle East, or than the former Roman Empire. The loss of the North African and Middle Eastern lands to the Islamic world caused the core area of the ancient Western world, the Mediterranean basin, to decline in importance relative to the marginal lands of Northern and Western Europe: the Mediterranean, once the center of one world, instead became the border between two. The center of gravity in Europe shifted north, and the population of Northern Europe overtook that of the Christian Mediterranean world around 800 AD.

The geopolitical configuration of the Western world changed from that of a single dominant core area, centered on the Mediterranean basin, into one with multiple core areas in Northern and Western Europe. The more fragmented geography of northwestern Europe gave it a smaller mass factor than the other major civilizations with the same military technology. Moreover, empire-building was disadvantaged by the prevalent military technology, and consequently the Carolingian Empire built by Charlemagne the Great could only be sustained under that man's exceptional leadership: the empire broke down quickly after his death.

Medieval Europe had two aspirants to universal imperial power: the Holy Roman Empire and the Papacy. Both however fell far short of being a truly imperial power by the standard of the classical Roman Empire. While the reign of the Charlemagne over his empire was too short, the rule of the Holy Roman Empire was too decentralized. Moreover, by the late medieval era, even these modest pretenders to imperial power were facing encroachment and undermining from the rising power of trading cities, regional princes and national states. Consequently, Europe had a highly pluralistic international order throughout most of the long medieval era.

Given the small medieval mass factor, the highly even distribution of capability in Europe generated a moderate level of relativist concern and developmental drive. Furthermore, given that the distribution of capability was quite even, the power-induced risk attitude was largely risk-neutral. The yoke of imperial conservatism due to a high concentration of power of the Roman era was lifted for Europe. The continued development of the Western legal tradition during the medieval era is important evidence of European vigor and drive during this era. First, there was the effort in the Byzantine Empire of Justinian's Code.[133] After that came the efforts made in Southern and Western Europe to revive Roman law, after the chaos of the Dark Ages had ended. There

---

[133] Refer to Tellegen-Couperus (1993) on the history of Roman law.

was also the development of canon law under the leadership of the Papacy and the autonomous development of merchant law and maritime law by the merchant community and the trading city-states. In fact, according to Berman (1983, 2003), the legal foundation of modern Europe and the modern world as a whole was laid down during the medieval era.[134]

The most advanced region of medieval Europe was Italy, partly because the Italian peninsula inherited much of the ancient Roman heritage. This heritage, in combination with the proximity to the Middle East and the Byzantine Empire, made Italy the forerunner in economic development in the Western world. Furthermore, northern Italy has a compact geography and was shielded from the imperial power of the Holy Roman Empire. The geographical barrier imposed by the Alps and the prestige and power of the Papacy neutralized the imperial power of the Holy Roman Empire in Italy and allowed the existence of a highly independent mini state system.

The rivalry between the Holy Roman Empire and the Papacy gave the city-states of Italy much de facto sovereignty; the Holy Roman Empire and the Papacy, as well as the Alps, in turn shielded Italy from the other great powers. This was the geopolitical basis of the mini city-state system of Italy.[135] The compactness of the northern Italian plain meant that even with medieval military technology, warfare and military contests between constituent political units were quite decisive and intense. That is to say, the medieval Italian city-state system with its compact geography had a larger mass factor than the greater European system which had a multiple core areas fragmented landscape.

The even distribution of capability in Italy and the greater mass factor produced a very high level of relativist concern. The result was the Italian premium cultural and economic position in medieval Europe and the splendid Italian Renaissance. In contrast, the more fragmented and compartmentalized geography of northwestern Europe dented the sharp edge of international political-military competition and diluted the drive for greater power and economic development.

Italian city-states engaged in intense rivalry against each other, as well as against external powers. The intense political-military competition generated a high development drive. Furthermore, given that the distribution of capability was highly even within Italy, there was no extreme power-induced risk attitude to distort decisions including economic one.

---

134 Refer to Blum and Dudley (2003) on how Latin as a lingua franca aided economic growth in medieval Europe.
135 There were mini city-state systems elsewhere, such as in the southern part of France. These other mini city-state systems were later subsumed under the nation states and did not last as long as in the Italian city-state system. Refer to Wesson (1978).

The commercial and naval rivalry between Venice and Genoa is famous: Venetian and Genoese navies dominated the Mediterranean Sea.[136] For instance, in the fifteenth century, Venice had some three thousand ships, ensuring its position as one of the great powers of Europe, and that Venetian navy was in fact the chief European bulwark against Turkish powers in the Mediterranean. The power of Venice is again indicated by its successful defense of sovereignty against a league of the Papacy, the Holy Roman Emperor, France, Spain, and some minor powers in 1508 AD.

These Italian city-states achieved an amazing level of prosperity besides being very productive culturally. Despite constant wars between the city-states, Italy remained the richest region in Europe: for example, the revenues of Florence in the fourteenth century were surpassed only by those of the king of France. It has been claimed that the cities of northern and central Italy alone possessed more wealth than the rest of the European continent combined.[137]

Collectively, the Italian city-states bestowed to the later world institutional innovations such as the civilian control of the military, marine premium insurance (which began in Genoa ca 1350 AD), and the double entry accounting system (which they learned from the Arabian merchants).[138] Of course, it is their cultural achievements, collectively termed the Renaissance, or "revival of learning", that are best known to us. Effectively, the European competitive state system, as well as the European Miracle, started in northern Italy. In a way, the relationship between Renaissance Italy and the European state system was analogous to the relationship between the city-state system of classical Greece and the succeeding Hellenistic era territorial state system. A mini state system ushered in a larger state system which was its intellectual inheritor. Splendid indeed was the vigor generated by the competitive rivalry among the city-states of Italy.

7. China

In China, like elsewhere, the advent of the medieval form of warfare reduced economies of scale in conflict. Since China itself during Qin, Han and Jin Dynasties was the dominant power in the Chinese international political system and commanded the lion's share of resources and capability within the system, the diminished mass factor brought forth a more even distribution of resources and

---

136 Refer to Wesson (1978, p. 159).
137 Refer to Wesson (1978, p. 158).
138 Refer to McNeill (1982) and Haywood (1997, p. 170).

military capability. The medieval military revolution also caused the settled agrarian Chinese society to enjoy lesser relative combined military and economic efficiency versus the nomads on horseback. The adverse shift in relative efficiency reinforced the effect of a smaller mass factor to dissolve the gigantic Chinese empire of the late classical era, and the result was a pluralistic international political system in the medieval Chinese world.

In China, the reign of the Han Dynasty ended around 220 AD. It was succeeded by the Period of the Three Kingdoms (220-280 AD), an era of civil wars. The indigenous Chinese Western Jin Dynasty united China in 280 AD, but soon fell victim to rampant intra-elite power struggles and exhaustive civil wars, and was unable to defend northern China against nomads within and beyond the borders. Under incessant nomadic assaults, the Jin Dynasty retreated to southern China in 317 AD. Until ca 581 AD, a series of five indigenous Chinese dynasties reigned in southern China: the Eastern Jin Dynasty (317-420 AD), the Song Dynasty (420-479 AD), the Chi Dynasty (479-502 AD), the Liang Dynasty (502-557 AD), and the Chen Dynasty (557-589 AD).

After the Jin Dynasty had retreated to the south, northern China plunged into complete chaos for over a century. Nomadic tribes and Chinese regional powers engaged each other in messy warfare. This period (308-439 AD) is known as the "Era of Sixteen Kingdoms by Five Tribes of Barbarians" in Chinese history. This is the Chinese counterpart of the Dark Age of Europe after the collapse of the Western Roman Empire. Stable governance only returned ca 439 AD to northern China under the authority of the Northern Wei Dynasty (386-534 AD), a regime formed by nomads. After the Northern Wei Dynasty, northern China was divided between, on one side, the Eastern Wei Dynasty (534-550 AD) and its successor, the Northern Chi Dynasty (550-577 AD); and on the other side, the Western Wei Dynasty (535-557 AD) and its successor, the Northern Zhou Dynasty (557-581 AD). The Sui Dynasty finally unified northern China as well as southern China, in the latter part of sixth century. However, the reign of Sui Dynasty was very short, lasting less than three decades. Sui Dynasty was soon replaced by the Tang Dynasty after a brief period of civil wars.

During this period of prolonged political fragmentation, there were many famous battles where small forces defeated opponents who were numerically superior. These victories attest to the smaller medieval mass factor which increased the chances of success of the smaller forces against bigger fish in military contexts. The more famous of these battles include the Battle of Guandu (200 AD), where Cao Cao defeated Yuan Shao; the Battle of Chi Pi (208 AD; the Battle of Chi Pi is also known as the Battle of Red Cliffs), where the alliance of Sun Quan and Liu Bei defeated Cao Cao; and the Battle of Fei Shui (383 AD), where the Chinese Eastern Jin Dynasty of southern China defeated the nomadic Qin Dynasty of northern China. The many battles won by Li Shi Ming (Emperor Tai Zong) of the Tang Dynasty were also victories for numerically inferior forces.

The change from the late classical Chinese empire to the medieval plural-istic Chinese international political system intensified political military com-petition and generated higher relativist concern despite the small medieval mass factor. It also freed China from the extremely risk-averse power-induced risk attitude of the late classical era and delivered China from the yoke of imperial conservatism and severe economic distortions due to an extreme risk-averseness. Consequently, creativity and developmental drive returned to China.

In this era of fragmentation, decentralization and instability, the grip of Confucianism on the Chinese mind loosened. A good example was the states-man Cao Cao, the de facto founder of the Kingdom of Wei of the Era of Three Kingdoms. Many of his policies were contradictory to Confucian teachings, and Chinese historians have classified him as a Legalist. Further evidence of the decline of Confucianism was the popularity of Naturalist Taoism among Chi-nese intellectuals. There was more individual creativity in the cultural arena as well: poetry enjoyed resurgence and literature showed a liberal tendency. Additionally, there were innovations in political institutions: these innovations produced the institutions which were inherited by the Tang Dynasty and cop-ied throughout East Asia, including in Japan.[139]

Under the threat of the nomadic regimes in northern China, the indigenous Chinese dynasties in control of southern China made many efforts to advance the economy. In the ancient era, China had one core region: the northern China Plain and the banks of the middle and lower stretches of the Yellow River. Due to the development of southern China from the third to sixth centuries, by the Sui and Tang Dynasties, the economic strength of the south became consider-able and had already caught up with that of northern China: another core area had thus arisen. The Sui Dynasty constructed the Grand Canal and tapped the economic resources of southern China for maintaining the imperial regime in the north. Henceforth, the two regions were merged into a single dominant economic core area.

Nomadic invasions and dynastic turnovers plagued medieval China. Even the famous Tang Dynasty achieved only a short unification of about a centu-ry and a half, unlike the Han Dynasties whose unified empires had lasted for more than four hundred years. The Tang Dynasty and the very brief Sui Dynasty were much weaker imperial regimes than the Qin and Han Dynasties, given the smaller medieval mass factor. That is to say, since medieval military conflicts were more indecisive and favored the stronger contestant less, the imperial power became less entrenched and secure, and faced more political-military

---

139 Refer to Wesson (1967, pp. 377-378).

challenges from within and beyond the border. The Tang Dynasty therefore had a higher relativist concern compared to the Han Dynasty.

The early Tang Dynasty enjoyed about a century of stability and prosperity due to a succession of three capable emperors: Emperor Tai Zong (who reigned from 627-649 AD), the de facto founder of the dynasty; Empress Wu (who reigned from 684-705 AD), the only female emperor in Chinese history; and Emperor Xuan Zong (who reigned from 712-756 AD). (In a way, their centralizing reigns, during a time more conducive towards decentralization, are analogous to Charlemagne's control of the Carolingian Empire: in both regions, only exceptionally skillful leaders could hold large empires together.) Consequently, notwithstanding some counterproductive policies of confiscation and government control and an underutilization of technology, there were considerable economic and cultural achievements during this era.[140] During this time, Tang China and the Islamic Arabian world were the most advanced regions of the world.

The Tang Dynasty produced a significant proportion of China's greatest poets and essayists; internal commerce and external trade thrived; the merchant class again enjoyed the freedom and high status they had been deprived of during the Qin, Han and Jin dynasties. The culture and political institutions of the Tang Dynasty spread to surrounding countries, including Japan. One of the two most important achievements in Chinese legal history, the Kai Huang Lu (the Kaihuang Code), compiled in the Sui Dynasty, laid down the basic legal institutional framework for China from the Tang Dynasty to the Qing Dynasty: a long-lived legal tradition indeed.[141]

Given the medieval military technology, centralized governance was hard to maintain, even in China. The reign of the central government of the Tang Dynasty lasted approximately a hundred years. The rebellions of An Lu Shan and Shi Si Ming from 755-761 AD ended the de facto control of the central government. China after the rebellion was politically fragmented in fact if not in name. Consequently, the Tang Dynasty failed to have an imperial rule that was long and secure enough to completely stop the momentum of growth and creativity that had accumulated during the earlier era of political fragmentation and competition.

More often than not, there was a de facto state system or pluralistic international order operating in medieval China, even if historical records often claim a dynasty was in charge. The contestants were of diverse origins: Chinese, Turkic, Tungusic, or Mongolian, or Tibetan or others. The more even distribution of capability intensified political-military competition, raised relativist concern,

---

140 Refer to Mo (1995; 2004, pp. 57-74) and Bernholz and Vaubel (2004, p. 5).
141 Refer to Head and Wang (2005).

reduced the power-induced risk averseness of the imperial regime and produced a high developmental drive that propelled China to first recover from the chaotic aftermath of collapse of Western Jin Dynasty rule in northern China, and then to reach the high achievements of the Tang Dynasty, Five Dynasties and Song Dynasty eras.

From the end of the rebellions of An Lu Shan and Shi Si Ming until the formal termination of the Tang Dynasty, economic and military power lay with regional commanders. These regional powers were practically kingdoms of their own. By the late ninth and early tenth centuries, wars and annexations had configured them into about eleven major states, one on the central plain of northern China and ten in other regions. This period was thereby named the Era of the Five Dynasties and Ten Kingdoms (the "five dynasties" refer to the series of powers in control of the northern China central plain and the "ten kingdoms" refer to the ten smaller states in other regions). Political fragmentation, competition, and economic development of this period continued into the succeeding Northern and Southern Song Dynasties and produced high levels of cultural economic achievements—levels that were not repeated in the China of either the Chinese Ming or Manchurian Qing era. These achievements are known as the Song Puzzle.

By 979 AD, the Song Dynasty, which originated in northern China, had conquered most of southern China. However, Yunnan province was in the hands of the Tali Kingdom, a tribal confederation of Thai peoples, and Vietnam achieved independence from China during the Era of Five Dynasties and Ten Kingdoms (907-960 AD). Another power of this state system was Tibet. Northern China was in the hands of three major powers: the Song Dynasty, which controlled the plain along the middle and lower stretches of the Yellow River; the Liao Kingdom (907-1125 AD), which controlled the Manchurian plain, inner Mongolia and parts of the northern China Plain; and the Xia Kingdom (1032-1227 AD), which controlled the Gansu Corridor and the upper stretch of the Yellow River valley. Of these three major powers, the Song Dynasty had the dominant share of the population and of economic and industrial might. However, the Liao and Xia Kingdoms, both highly sinicized semi-nomadic and semi-agrarian societies, together almost monopolized a very important strategic item: horses. Hence although Song China had greater economic might, relative combined military and economic efficiency among the three major powers were quite even and so was the distribution of military capability.

The Xia and Liao Kingdoms were unlike the nomads who constantly harassed the borders of China. The nomadic tribes lacked the resources to really conquer China, the cultural sophistication to administer China, or the economic and fiscal efficiency to engage China in prolonged, large-scale military struggle. The Xia and Liao kingdoms, besides having strong military power due to the possession of horses and skilled horsemen, also controlled a sizable amount of resources, commercial networks and industrial capacity. They also had the support of bulky and strong state machinery. Given their control of large tracts of settled territory, these two semi-nomadic semi-agrarian states were more akin to the earlier Northern Dynasties of 386-581 AD, which had ruled over northern China and posed a serious threat to the indigenous Chinese regime in the south. Established by nomadic tribes which had successfully assimilated Chinese culture and statecraft, the Northern Dynasties proved to be more than the political and military match of the Chinese regime in the south and their political achievements laid the foundation for the unifications led by the northern powers of the Sui and Tang Dynasties.

If the fact that the semi-nomadic states of Xia, Liao and Jin made up for their inferiority in economic resources with superior military efficiency is ignored, Song China would be classified as another example of an imperial power commanding a preponderant share of capability. The truth however is that Song Dynasty China was more like the bipolar era of the Northern and Southern dynasties, rather than similar to the short-lived imperial orders of the Sui and Tang dynasties. The Song Dynasty was even more unlike the stable and enduring imperial orders of the Qin Dynasty and Han Dynasty or, the Yuan, Ming and Qing dynasties that came after.

For all that the Song may have claimed dynastic control over the whole of China, the three major states of Song-era China treated each other as worthy rivals. In fact, Song Dynasty often had to pay tribute to her northern rivals to maintain peaceful relationships.[142] Of the three major states, the Northern Song and Liao were the most powerful players. The Liao Kingdom's territorial size was twice that of the Northern Song. The Jurchen Jin Kingdom eliminated the Liao Kingdom in 1125 AD and occupied Liao territory, as well as annexing the middle and lower stretches of the Yellow River region from the Song Dynasty. Militarily the Liao and Jin Kingdoms were usually on the offensive whilst the Song Dynasty was on the defensive.

After 1127 AD, the Song Dynasty, which retreated to the southern part of China, and the Jin Kingdom (1115-1234 AD) were about equally matched. Song-era China was therefore a tri-polar state system. The economic powerhouse of the Song Dynasty was southern China, especially the middle and lower stretch of the Yangtze River basin. The loss of the northern China central plain by the Southern Song Dynasty therefore was not a significant blow economically.

During the Song Era, due to the pressure from the northern rival states, there were many military technological innovations. To meet with the threats from powerful rivals, the Song Dynasty put significant effort into researching the use

---

142 Refer to Lee (1988).

of gunpowder in warfare. Other innovations in military technology included a greater use of mechanical missile devices which increased the power of the infantry (during the Northern Song Dynasty). There were widespread uses of gunpowder weaponry during the Southern Song Dynasty. Song China and its rivals maintained large-scale standing armies and navies. The large scale of the defense forces in the Chinese competitive state system was greater than that seen in Europe until the Napoleonic wars.[143] Keenly aware of its military disadvantage due to the short supply of horses, Song China strove to at least partially make it up through greater economic might and technological superiority. An extensive network of industrial-military development was maintained to tap economic resources and technological prowess for military use.

Unlike the decentralized Tang Dynasty, Song China was a very centralized service state which intervened extensively to help advance the economy. Song China produced many capable statesmen, of which the great reformer Wang An-Shih (1021-1086 AD) was the most famous. Wang's reform programs aimed to create a service state for greater economic might and military prowess, all in order to meet the challenge posed by the northern rival states. Wang also tried to trim down the over-sized state sector and increase its efficiency. State civil employees numbered in the millions during the Song era, as did military personnel. Wang's reform effort was from 1069-1085 AD, when he was the prime minister. Wang, however, was neither the first nor the last reform-minded prime minister of the Song Dynasty.

On the legal-institution front Song China was innovative without being revolutionary. The Song Dynasty retained the basic legal framework inherited from the Tang Dynasty and made extensive use of imperial edicts to handle new economic realities and changes. These imperial edicts were administrative orders that, in essence, replaced the Song Codes which followed the legal framework of Tang Codes and the Kai Huang Lu of Sui Dynasty. Such flexible and expedient measures allowed the legal institutions of the Song Dynasty to cater to the new commercial and industrial sectors without challenging the fundamentals of the traditional legal system.[144]

The economy prospered with elaborate internal and external trading networks.[145] The members of the global trading network of this period included Southeast Asia, India, the Middle East and Europe. External trade brought in close to half of the state revenue: the economy was so commercialized that soon after 1000 AD the Song government found it more convenient to collect taxes in cash instead of in kind. There was sustained growth in per capita

---

143 Refer to McNeill (1982) and Keegan (1993).
144 Refer to Head and Wang (2005, p. 145).
145 Refer to Abu-Lughod (1989) for the global trading network at this period.

income during this era and China was the industrial center of the global trading system. Jones (1990) notes that:

> "Under the Song, at any rate, there is evidence of: a) structural change (when labor moved from primary production to higher-yielding secondary or tertiary production; in the modern world this is positively associated with rising per capita GNP); b) the large-scale purchase of consumer goods; c) and widespread, quite advanced, technical innovation." (7)

Song China led the world in shipbuilding and maritime technology. Song China's innovations include the invention of the compass, the adjustable centerboard keel, and the use of cotton sails in place of bamboo slats.[146] This maritime technology and know-how was used in naval warfare between the Southern Song Dynasty, the Jin Dynasty and the Mongolian regime. These were large-scale engagements with hundreds, and at times thousands, of seagoing ships employed. Song China explored parts of maritime Southeast Asia, including Taiwan and the Philippines. The maritime achievements of the Song Dynasty laid the foundation for the later and perhaps more famous overseas expeditions undertaken by the famous eunuch Admiral Zheng He and others during the early Ming Dynasty.

During this period, China experienced a burst in technological and economic progress.[147] For instance, Song China invented and disseminated a water-powered hemp-spinning machine, whereas similar devices in Europe only appeared around 1700 AD.[148] By the end of the 11th century, China's total iron output had seen a peak of 150,000 tons; total European output only matched these levels some six centuries later.[149] Effectively, Chinese iron output per capita at the close of the 11th century was some twenty per cents greater than that of Europe. The invention of the unmovable printing press during the late Tang Dynasty and the invention of the movable printing press during the Southern Song Dynasty greatly facilitated the spread of information and led to greater efficiency in public administration as well as higher economic productivity. Three of the four greatest inventions of China were made during the Song Dynasty: namely, gunpowder, printing, and the compass. (The fourth, paper, had been invented earlier, during the Eastern Han Dynasty.) The superior economic performance of Song China is also testified to by the fact that the Song Dynasty was the only

---

146 Refer to Stavrianos (1982, p. 283).
147 Refer to Cipolla (1967, pp. 101-2) and Graham (1973).
148 Refer to Elvin (1973). In later periods China however actually abandoned the use of this machine and regressed in terms of industrial technology. Refer to Jones (1981, p. 64).
149 Refer to Hartwell (1966) and Harrison (1972, p. 290).

major Chinese dynasty that had no large-scale peasant rebellion. There were also major changes in land ownership and educational reorganization.

The Song Dynasty's cultural achievements were equally impressive: the porcelain manufactured during the Song Dynasty was the best in quality and design; the most important works of Neo-Confucianism were completed; poetry and other literary works flourished, with the comprehensive historical works of *Chi Zhi Tong Jian*, by Shi Ma Guang, appearing.[150] Such a miraculous performance was not repeated in the subsequent powerful and unified Ming and Qing periods.[151]

Commenting on the intensive growth of Song China, Stavrianos (1982) notes:

"In addition to its cultural attainments, the Song period is noteworthy for a commercial revolution with much significance for all Eurasia. The roots are to be found in a marked increase in the productivity of China's economy. Steady technological improvements raised the output of the traditional industries. Agriculture likewise was stimulated by the introduction of a quickly maturing strain of rice that allowed two crops to be grown each season where only one had been possible before. Also new water-control projects undertaken by the Song greatly expanded the acreage of irrigated paddy fields. Thus it is estimated that the rice crop doubled between the eleventh and twelfth centuries. Increasing productivity made possible a corresponding increase in population, which in turn further stimulated production in circular fashion. The volume of trade also rose with the quickening tempo of economic activity. For the first time there appeared in China large cities that were primarily commercial rather than administrative centers." (282-283)

Stavrainos (1982) further observes that Song China was the first government to introduce paper money, which facilitated commerce. Song China exported mostly manufactured goods, including silk, porcelain, books and paintings, and imported mostly raw materials such as spices, minerals and horses. This trade pattern testified to Song China's global economic leadership.[152]

Bernholz and Vaubel (2004) comment that:

"Under the Song dynasty (960-1275), which faced severe international competition and had to pay tribute to neighbouring 'barbarians' for most of its

---

150 Refer to McNeill (1999, p. 253).
151 Refer to Jones (1981, p. 202) and Bernholz and Vaubel (2004, p. 6)
152 Refer to Stavrainos (1982, pp. 283-4). Refer to Deng (2000) for a survey of works on Chinese economic history. Refer to Wesson (1967, pp. 204-205) for the developmental achievements of Song China.

reign, China advanced to the threshold of a systematic experimental investigation of nature and created the world's earliest mechanized industry. Private printing was popular, and water-powered machines for spinning hemp thread appeared. The government promoted education, irrigation and trade, including maritime and foreign trade. Military technology was at a high level. The Song period was both the climax and the end of scientific technological progress." (5)

The experience of medieval China, from the collapse of an indigenous Chinese regime in northern China during the Jin Dynasty to the end of the Southern Song Dynasty, was largely one of political fragmentation and competition as well as intellectual creativity and economic development. It largely accords with the main arguments of this book. Of special importance were the innovations of Song China in military technology which sowed the seed for the next major round of military technological change, that of the gunpowder military revolution.

Given the intense political-military competition of the Chinese system during the Song era, Song China led the world in military technological innovations.[153] Around 900 AD, China began to undergo major changes in military technology. The use of gunpowder-based weapons proved to be especially revolutionary. The destructive power of gunpowder was soon harnessed to bring down fortresses and other defenses, and this invention gradually spread to the other major cultures of the old world, including Europe, through the Mongolian empire. This was the beginning of the Gunpowder Military Revolution which, along with the concomitant innovations of the printing press and compass, began to fundamentally change the world.[154]

The Mongolian military combined the battlefield mobility of light cavalry with the military technological breakthroughs generated by the intensely competitive and highly innovative Chinese international political system to conquer a large part of the Eurasian land mass. The Mongolian Empire had greater war efficiency and a larger pool of resources at its disposal than the earlier northern rivals of Song China. Relative combined military and economic efficiency had become more asymmetric. The economies of scale in warfare had become greater too, due to the use of gunpowder, especially in overcoming fortresses and other defenses. Consequently, there was a greater concentration of resources and capability in the hand of the leading political-military power. The Mongolian Empire therefore embarked upon a path of far-flung conquests

---

153 Refer to McNeill (1982) and McNeill (1999, Ch. 14). McNeill (1982) refers to the period from B. C. 200 to A. D. 1200 as the millennium of Chinese leadership in military technology.
154 Refer to Needham, Ho, Lu and Wang (1987).

hitherto unknown in history. The Mongolian Empire then used the resources so acquired to overrun the Southern Song Dynasty. The conquest of Southern Song Dynasty by the Mongolian Empire ended the medieval fragmentation of China and with it the Song Puzzle.

8. Japan

The medieval era saw the entrance of Japan into the civilized world. By the end of the 4th century, the rulers of the Yamato plain in Honshu had created an extensive kingdom. During the 6th century, Prince Shotoku further transformed the Yamato Kingdom by strengthening the authority of the court over the provincial nobles and building an administration based on the Chinese model.[155] The influence of Tang China was extensive. Japan adopted the Chinese model of military and central government. Japan also borrowed the Chinese writing system and other aspects of Chinese culture, such as Confucianism and Taoism.[156] This period was referred to as the Taka era of sinicization.

The centralization scheme succeeded only partially at most. The central state in Japan was never able to consolidate its authority. Japan has a highly fragmented geography: the four major islands are very mountainous, and mountain ranges dissect these islands into many compartmentalized small coastal plain core areas. Due to the highly fragmented geography which aided peasants' resistance, central taxation of the agricultural sector was not effective; public administration devolved into decentralized local units. There arose the large estates which formed the economic foundation of feudalism, which lasted from around the 11th to the 15th century.[157]

The trend toward decentralization and fragmentation was reinforced by the introduction of horses and cavalry warfare from the Eurasian mainland. Given the medieval military technology, the Japanese system had small economies of scale in warfare. Consequently, there was a highly even distribution of resources and capability among the many contesting units. Feudalism and warlord politics replaced centralized government. In place of the imperial court, the military leadership of the shogunate exercised effective rule, though not in name (power was of course officially held by the emperor). Yet, by the late medieval period (ca 1150-1200 AD), even the shogunate and feudal lords found it hard to control the society, particularly the peasantry.

---

155 Refer to Haywood (1997, pp. 112-113).
156 Refer to Raaflaub and Rosenstein (1999, pp. 52-56).
157 Refer to Haywood (1997, pp. 228-229).

From the onset of the daimyo (ca 1330 AD), but prior to the unification of the country under the Nobunaga, Hideyoshi and Tokugawa Shogunates, there were several hundred feudal units in Japan. Given the highly even distribution of capability, even with the small factor due to the medieval form of warfare, there was a significant amount of relativist concern generated. Moreover, there was no highly risk-averse power-induced risk attitude to distort economic decisions or hold back creativity given the highly even distribution of capability. Consequently, it was a period of rapid economic growth with important technological advances in agriculture and handicrafts. There was the transition from the barter economy to the money economy, with a significant increase in foreign trade. Japanese traders were active throughout Southeast Asia by the late fourteenth century.[158]

The succeeding period of 1467-1573 AD, one of intense military contests for power, is known as the Era of Warring States in the history of Japan. Firearms were imported from the West, especially through Portuguese and Dutch traders: the gunpowder military revolution thus affected Japan. The gunpowder military revolution raised the economies of scale in warfare of the Japanese system. There were heightened political-military competitions between the warlords. Relativist concern was elevated to a new height and generated hectic economic developments among the Warring States of Japan. There were extensive irrigation works, mining, trade, highway construction, fiscal capacity development and rapid urbanization. The intense political-military competition among the many warring states ended only with the ultimate unification of Japan under the Tokugawa Shogunate.

9. Eurasia

The mega Eurasian international political system had taken a more concrete shape during the medieval era. The dominance of light and heavy cavalry in battlefields gave medieval military forces great mobility and projection power through geographical space. Consequently, political and military competition among the major Eurasian cultures increased. Parallel to the development of greater overland projection power of the medieval military forces was the formation of a Pan Eurasian maritime trading network that knitted together all the major regions: Europe, the Middle East, South Asia, Southeast Asia and East Asia.[159] The Islamic civilization was the linchpin in both the medieval mega

---

158 Refer to Stavrianos (1982, pp. 292-293).
159 Refer to Abu-Lughod (1989).

Eurasian international political system and the medieval pan Eurasian mari-time trading system. Therefore, the medieval major cultures were keenly aware of the existence of each other and were in constant cultural and economic and even political and military contacts.[160]

Despite the frequent contacts, none of the interactions between the major cultures had a political-military nature that was frequent or serious enough to generate sustained and significant relativist concern, with the exception of the political-military competition between the Middle Eastern based Islamic empires and states and the Byzantine Empire and European states. Consequently, political competition *between* (rather than *within*) the major cultures was not a major force for generating sustained and significant development effort. In other words, though the Eurasian mega international political system and mega international economic system took a more definite shape in the medieval time than during the late classical era, the mass factor between the major Eurasian cultures was still too small during the medieval era to spur substantial developmental drive.

Of all the relationships between the major cultures, that between China and the Middle East had gone through the most drastic change from the late classical era to the medieval period. Unlike the Chinese-Persian relationship of the late classical era, which a complete non-rivalry, there was for a brief period an acute political-military competition between China and Islam over the control of Central Asia. At the peak of its power, the Tang Dynasty of China established military bases in Central Asia for the protection of the Silk Road, something that China had not done for a long time—not since the collapse of the Han Dynasty. Yet, in the Battle of Talas (751 AD), the forces of China's Tang Dynasty were defeated by the Arabian army of the newly-founded Abbasid Caliphate. This defeat, together with the internal turmoil started by the Rebellions of An Lu Shan and Shi Si Ming, prompted China to retreat from Central Asia. Since then the two major cultures had no significant political-military competition given the great distance between them. So although China and the Middle East had little direct rivalry at the beginning or at the end of this period, there was for a brief time intense competition over Central Asia.

The political-military relationship between China and India during the medieval era was very different from that which existed between China and Islam. Medieval India was politically much more fragmented than China. Therefore, it was a one-way exertion of political-military influence by China on India when China was strong: India itself had little political military influence over China. At the peak of its power, the Tang Dynasty of China established

---

160 Refer to Frank and Gills (1993) and Franks (1998).

military outposts in northern India to protect its interests. But all these ended with the defeat of Tang forces at the hands of the Arabian army in the Battle of Talas. In sum, the almost impassable Tibetan Plateau and the Himalayan Ranges effectively isolated and buffered China from direct political military competition with the two adjacent major cultures of Islam and India before the modern era.

During the medieval era, Islamic forces made constant incursions into the politically fragmented medieval Indian state system. There was no Tibetan Plateau or Himalayan Mountains to buffer or isolate India from the constant encroachments of Islam. There were many conversions to Islam, mostly from the lower castes in the Hindu society who were attracted by the egalitarian teachings of Islam. The political-military competition among the Hindu states and with the Islamic forces generated some sustained development, especially in southern India.

Other than China and the Middle East, Southeast Asia was the major region closest to India. However, mainland Southeast Asia was separated from India by geographical barriers of torrential rivers, impenetrable forests and high mountains. The Indian Ocean and Bengal Bay were less hostile but nonetheless significant barriers between the South Asian subcontinent and maritime Southeast Asia. Due to the vast distance and rough terrain between them, there was no significant sustained political-military competition between the two major regions during the medieval era. A rare exception was in the eleventh century, when the southern Indian maritime Chola Empire had naval expeditions to Southeast Asia and clashed with the Srivijaya Empire which was based in southeastern Sumatra. Though it severely undermined Srivijayan hegemony, the Chola invasion was ultimately unsuccessful.

The most intense political-military competition between the major cultures was that between Islam and the West. The Islamic Caliphate expanded into Central Asia, South Asia and Southern Europe, and the threat from Islam is often cited as one of the reasons for the rise of the West. Islamic advancement generated significant relativist concern in the Western world, especially in the Byzantine court and the Christian principalities bordering the Islamic world. The series of Crusades in the 11th, 12th and 13th centuries, by Christian armies trying to retake Palestine from the Muslims, was one of the consequences of such concern. Partly due to the competition and contact with the Islamic world, Southern and Mediterranean Europe was the most advanced part of the Western world during the medieval era and Spain and Portugal spearheaded the great geographical discoveries and the European march towards world leadership.

Japan was probably the most geopolitically isolated major Eurasian culture during the medieval era. Situated on a string of islands located at the far eastern end of the Eurasian landmass, Japan was geopolitically isolated from practically all the other major Eurasian cultures except China. The great width of the East China Sea and the Japan Sea also almost ruled out persistent serious political-military competition between Japan and China, the other major culture that was closest geographically. During the medieval era, Tang Dynasty China fought with Japan in Korea in the Battle of Baekgang (663 AD): Japan was defeated and retreated, but the Tang Dynasty however was in turn driven out of Korea by their erstwhile Korean ally about a decade later. After this conflict, the two major cultures had no military contacts for centuries.

The Mongolian Empire was the first truly pan-Eurasian power. It had direct military contacts and competition with practically all the major Eurasian cultures: its military campaigns reached Europe, Japan, South Asia, Southeast Asia and the Middle East. Despite striking fear into the hearts of many across Eurasia, many of these ambitious far-flung campaigns either failed disastrously or ended inconclusively, due to vast distances and difficult terrain and weather conditions, despite the advanced weaponry and dominant resources of the Mongolian Empire. The empire was divided into four major realms for administrative purposes because of difficulties in communication and control over such vast distances, but despite such measures soon the empire broke down and dissolved. In other words, the mass factor was still too small on a pan-Eurasian level to sustain a stable pan-Eurasian empire or to generate significant and persistent political-military competition among the major cultures under normal circumstances, given the pre-industrial technology.

10. Conclusions

The use of heavy cavalry in the medieval era made cavalry the dominant force in battlefield and reduced the economies of scale in warfare. The smaller mass factor brought about massive structural changes in geopolitics. Continental size empires of the late classical era either retreated or dissolved. Political fragmentation was the norm in the long medieval era. Due to this fragmentation in the geopolitical landscape, during the medieval era, all major Eurasian civilizations exhibited significant though not miraculous creativity. Among the medieval bursts of creativity, China and the Middle East stood out and produced the most outstanding cultural and economic achievements, known respectively as the Song Puzzle and the Abbasid Golden Age. Yet the medieval Chinese and the early Islamic international political systems were hardly good examples of the state system. These achievements therefore seem to contradict the Hume-Kant Hypothesis and present difficulties for Bernholz and Vaubel (2004). The argument proposed by this book, with its more general theoretical underpinning, has no problem in dealing with these two cases as well as accounting for the developmental performance among the major cultures during the medieval era. That is to say, the experience of the medieval era agrees with and lends further support to the main argument of this book that military technology shapes the geopolitical landscape and that political military competition leads to cultural and economic creativity.

CHAPTER 7

The Rise of the West

1. Introduction

The early modern era started with an explosion of the gunpowder weaponry. The destructive power of gunpowder made warfare more lethal and raised the economies of warfare. As warfare became more decisive again, massive pan-continental-size gunpowder empires replaced the fragmented geopolitical landscape of the medieval era. Another important technological change that characterized this era was the great advance in maritime technology and geographical knowledge. The world was on the eve of great geographical explorations and discoveries and the making of a truly global international political system. At the beginning of the early modern era, it was not clear which major Eurasian civilization will break free from the rough equilibrium among the major Eurasian cultures to forge ahead and become the leader in knitting the different regions of the globe together into a global system. In fact, why ultimately it was Europe which emerged as the victor in this race among the major civilizations is still puzzling scholars.

The chapter especially focuses on the splendid development of Europe and its sharp contrast with the dismaying performance of the other major Eurasian cultures. The European miraculous success and the contrasting surprising failures of the other major Eurasian cultures are what motivated the formulation of the original Hume-Kant hypothesis in Bernholz et al. (1998) and Bernholz (2004) as well as many other investigations, including that of North (1981, 1987, 1990, 1995, 1998) and Jones (1974, 1981, 1988, 1990, 2002).

In Europe, the large mass factor brought forth by the gunpowder military revolution together with a fairly even distribution of capability among the major contestants generated higher relativist concern. Contestants increased their state capacity competitively, leading to the rise of centralized territorial national states and great power politics, characteristics associated with the Westphalian system. In the other major Eurasian civilizations, however, the large mass factor brought forth by the gunpowder military revolution resulted in extreme concentrations of capability. Continental size empires reemerged on the geopolitical landscape. Powerful and long enduring imperial orders were reconstituted. There were lower relativist concern and hierarchical international orders (such as the Chinese tribute system) in these diverse lands.

Japan is of special interest here. The Hume-Kant hypothesis of Bernholz et al. (2004) is unable to explain the moderately good economic performance of Japan under the unity of the Tokugawa regime. The Tokugawa Shogunate was another gunpowder empire. According to the Hume-Kant hypothesis, one would expect Tokugawa Japan to have performance similar to the other major Eurasian gunpowder empires of China, India, Persia and the Ottoman Middle East. However, the performance of Japan during the early modern era, though not as splendid as that of Europe, was above those of the other major gunpowder empires. The intermediate performance of Japan under the Tokugawa Shogunate could be explained if the effects of military decisiveness and distribution of capability within the Japanese system are taken into account.

Among all these diverse international political systems around the globe in the early modern era, only Europe had the propitious combination of a large mass factor brought about by the gunpowder military revolution and an even distribution of capability among the major contestants preserved by the fragmented multiple core area geography. The results were intense competition among the contestant for power, keen concern for balance of power, high relativist concern and, persistent and constant great effort to increase state capacity, economic might and military prowess.

2. The Gunpowder Military Revolution

Gunpowder was first discovered in Tang Dynasty China and was soon put to military use. Gunpowder was employed with devices such as rockets, primitive flamethrowers, and grenades launched from catapults. In particular, Song Dynasty China made regular use of cannon from the twelfth century onwards, first using bamboo cannons and then shifting to bronze cannons. The Mongolian tribesmen made excellent military use of gunpowder to conquer a far-flung pan-Eurasian Empire. Indirectly, economic exchanges among the major cultures were facilitated through the thriving intercontinental trade under the protection of Pax Mongolica and technology was inevitably one of the goods exchanged.[161]

Either through trade facilitated by Pax Mongolica or through wars, the technological advances associated with the use of gunpowder spread from the Mongolian Empire to the other major civilizations of the old world. Consequently, the gunpowder military technology transformed the geopolitical landscape of the various major Eurasian cultures, ended the fragmented medieval international orders and brought in larger political units such as the gunpowder empires

---

161 Refer to Anderson and Marcouiller (2005) for a theoretical treatment. Refer to Abu-Lughod, J. L. (1989) and Findlay and O'Rourke (2007) on the impacts of the Mongolian Empire on global economic exchanges.

and national states.[162] With the greater economies of scale in conflict created by the gunpowder military revolution, stable and hard to contest imperial orders returned to the old world for the first time since the collapse of the Roman, Gupta and Han empires almost a millennium earlier. In Europe, the Middle East, South Asia and East Asia, there were downward trends in the number of states and upward trends in the size of political units, caused by the greater economies of scale in warfare brought forth by the gunpowder military revolution.[163] This process began around 1200 AD and gathered momentum around 1400 AD.

The military use of gunpowder changed warfare in many ways. Gunpowder militarily forced advancements in not just weaponry, but also in offensive and defensive tactics, troop formation, logistics and military doctrines. Gigantic fleet navies were formed; mass infantry dominated armies and replaced the medieval cavalry-centered military. The most immediate and important effect of gunpowder was felt in siege warfare, as the use of cannons reduced the defensive effectiveness of castles and fortresses. The speedy downfall of Constantinople—whose massive defensive walls which had shielded the Byzantine Empire and Europe from invaders from Asia for over a thousand years—in 1453 AD, under the artillery bombardment of the besieging Ottoman Turkish army, signified the dawn of the modern form of warfare centered on gunpowder. Consequently, defensive walls were made lower and thicker and lined with cannons. Defensive networks became larger and more expensive and siege warfare increased in scale.[164]

In the field, first the deployment of field cannons and then the use of personal firearms such as arquebuses, muskets and rifles made infantry the king of the battlefield again. The use of new tactics such as volley fire, first used by Oda Nobunaga of Japan in the Battle of Nagashino (1575) in which Nobunaga's arquebuses armed infantry delivered a decisive defeat to opponent's charging cavalry, and the Dutch in Europe a couple decades later, eventually rendered cavalry almost useless except for its use in reconnaissance. From then on it was the amount of firepower that an army could put into fighting that determined the outcome of warfare, as a larger force commanding greater fire power could easily overwhelm a smaller foe swiftly and decisively. Land battles became decisive again. Field battles grew larger in terms of both manpower and firepower. Large standing armies were raised to replace medieval cavalry, whilst at seas, the fitting of cannons to ships transformed naval warfare and made naval battles more lethal and decisive.

---

162 Refer to Stavrianos (1982, pp. 247, 254). For the early use of gunpowder in Europe and the Islamic world, refer to Goldschmidt (2002, Ch. 9).
163 Refer to Parker (1996).
164 Refer to Parker (1996).

In sum, the gunpowder military revolution made siege and field battles more decisive. In other words, it increased the economies of scale of warfare. There were increases in the size of the military and greater demands for manpower and resources. These larger financial and logistical needs in turn stimulated the formation of a huge, specialized and professional bureaucracy to service, support and control the military. Consequently, modern state and modern military replaced feudalism and its military system. Feudalistic small principalities and kingdoms gave way to large centralized territorial states and empires better able to exploit the economies of scale in warfare and able to finance the expensive war machines. There arose the gunpowder empires of the oriental world and the national states of early modern Europe.

3. The Gunpowder Empires

The gunpowder military revolution increased the economies of scale in warfare. Given an asymmetric relative combined military and economic efficiency among the contestants, a greater mass factor led to higher concentration in the distribution of resources and capability in the hand of the stronger contestant. The impact of these changes on China will be examined first, as the gunpowder military revolution started there.

China had a single dominant core area: the plains along the middle and lower stretches of the Yellow River and the Yangtze River linked up by the Grand Canal. This dominant core area contained the lion's share of China's population, industrial capacity and economic resources. Consequently, China had a highly asymmetric relative combined military and economic efficiency, as a political contestant occupying this dominant core area enjoyed overwhelming advantages in taxation, defense and other administrative and economic aspects against other contestants outside the dominant core area. The gunpowder military revolution therefore led to an extraordinarily high concentration of resources and capability within the Chinese system. It ended the pluralistic and contestable international political order of medieval China.

It was the Mongolian tribesmen with their excellent equestrian skills that initiated these mega changes in geopolitics. They were the first to extensively exploit the power of gunpowder for imperial pursuits. The Mongolians, being nomads who specialized on mounted warfare, largely used gunpowder in siege battles to bring down fortresses and had not fully exploited the potential of gunpowder in field battles. But the use of gunpowder weaponry in siege warfare alone was enough to confer great military advantage to the Mongols. Combining the new destructive power of gunpowder with tribal coherence and the mobility of the steppe people, the Mongolians enjoyed an extremely formidable advantage in relative combined military and economic efficiency against all other contestants. The outcome of this combination of extreme asymmetry in state-building efficiency and a large mass factor due to the gunpowder military revolution was an empire larger than any previously known in the history of Eurasia and humanity.

The major states in the Chinese system of the time, Xia, Jin and Song China, were successively conquered by the Mongols, as were Tibet and the Tali Kingdom. Central Asia and Persia were captured early. In the Middle East, the Mongolian troops wiped out the Abbasid Caliphate, but the Egyptian mamluks defeated the Mongols at the Sinai Peninsula in the Battle of Ain Jalut (1260 AD) and stopped their advances. In Europe, the Mongolian Empire reached the Danube; in South Asia, Mongolian rule was established in Kashmir. The Mongolian Empire was the first pan-Eurasian empire and is the largest contiguous empire in the history of the world. Its emergence signified the beginning of more substantial political and military interactions among the major cultures on the pan-Eurasian level.

China was the first major Eurasian culture to enter the era of gunpowder and the first major culture to have a continental-size gunpowder empire. The Mongolian tribes combined the technological achievements of the military revolution of the Song China competitive state system with the mobility of the nomadic people to conquer the Chinese world and a world empire larger than all before. However, the economies of scale in the military aspect were not matched by the economies of scale in public administration. The Mongolian empire soon split into four successor empires which were separately ruled.

The Chinese part was known as the Yuan Dynasty, and it reigned for nearly a century (1279-1368 AD). It was succeeded by the indigenous Chinese Ming Dynasty, which governed for almost three centuries (1368-1644 AD). The Manchurian Qing Dynasty in turn replaced the Ming Dynasty and ruled for about three centuries (1644-1911 AD). Given the large mass factor and the high asymmetry in relative combined military and economic efficiency, the imperial order in the Chinese system was highly stable. Despite two dynastic turnovers and several major civil wars, the imperial order was reconstituted swiftly again and again. The lion's share of resources owned by the single dominant core area ensured that any contestant controlling it enjoyed overwhelming odds of victory against any rival. This imperial order lasted for over seven hundred years and was overthrown only when China was threatened and encroached upon by industrialized Western great powers. The First Opium War of 1839-1842 AD signified the induction of China into the modern world state system, and the centuries-old imperial order of China was thus ended when China was incorporated into the Europe-centered modern world state system around 1850 AD. The Middle Kingdom's geopolitical isolation ended, and from then on, China had to reckon seriously with the military threat of states other than those in its vicinity.

Given the large mass factor of the gunpowder military technology and the extremely high concentration of resources in the Chinese system, relativist concern was extremely low in comparison to medieval China. Furthermore, given the extremely high concentration of capability and the large mass factor, the power-induced risk attitude was very risk-averse. The extremely risk-averse attitude caused severe distortions in the economic decisions of the imperial regime. Together, the extremely low relativist concern and extremely risk-averse attitude and the consequent imperial complacency and conservatism led to a very drastic change in the fate of the Chinese economy and society, and a stark difference between the developmental achievements of medieval China, as attested by the Song Puzzle, and the developmental performance of China under the three consecutive gunpowder dynasties.

During the Mongolian Yuan Dynasty, the effects of imperial stagnation had yet to come into full play. The economic and cultural momentum of the medieval era and its pluralistic and competitive international order had not yet completely run out: China was still a place of economic prosperity and cultural creativity.[165] However, by the time the Ming Dynasty was established, that momentum had practically worn out. This downward trend was slowed, but not reversed, under the Manchurian Qing Dynasty. The Manchurian rulers were more on the alert and hard-working since they were governing a subject population of a different race and culture with far greater number and cultural sophistication. Nevertheless, the Manchurians inherited the Ming policies and mind-set.

The Mongolian Yuan Dynasty used only coercion to maintain rule. Their brutal but crude statecraft allowed cultural and economic vigor to maintain momentum. The merchant class retained their freedom and high status, whilst the Confucian literati were designated to a very low social standing: a situation which actually helped to free Chinese minds. However, once under the more perfected imperial rule of the indigenous Chinese Ming Dynasty and the heavily sinicized Manchurian Qing Dynasty, Chinese cultural and economic vigor subsided and largely disappeared.

The Ming and Qing Dynasties effectively used the imperial examination system to control Chinese minds. Scholars aspiring to a career in officialdom were forced to self-indoctrinate in Confucian teachings; the imperial examinations prescribed the use of eight-legged essays which emphasized formality and memorization in place of independent critical and creative thinking. On top of this, the imperial regime also prescribed official interpretation of Confucian texts in imperial examinations. Furthermore, during the Ming and Qing

---

165 Jones (1981, p. 160): "China came within a hair's breadth of industrializing in the fourteenth century."

Dynasties, the imperial regime sponsored monumental cultural tasks such as the publishing of encyclopedias, and these tasks helped channel the intellectual energy of the Chinese literati to uncreative works that posed no threat to the imperial regime. The imperial regime also made use of such opportunities to scan the vast amounts of literature, erasing and editing any documents deemed unfriendly to the imperial regime. These measures were backed up by the use of brutal political reprisal in response to publication of material even remotely unfriendly to the regime. Effectively, what modicum of intellectual curiosity or creativity that remained in the Chinese mind was crushed.[166]

Under the indigenous Chinese Ming Dynasty, China retreated into isolation.[167] The Ming Dynasty severely restricted and eventually banned both international contacts and trade. After initially expanding their navy, the Ming court subsequently allowed that to decay, retreating from the sea as the leaders became more inward-looking.[168] The famous overseas expeditions of the Eunuch Admiral Zheng He were the last manifestation of the Chinese vigor. After the brief ambitious overseas expeditions of Admiral Zheng He, the Ming court decided in 1480 not to continue Chinese maritime explorations. The Confucian literati, who were against overseas expeditions and foreign trade, defeated the eunuchs who were in favor of it. Bernholz and Vaubel (2004) comment that:

> "After the rule of the Mongols from 1276 to 1367, the Chinese Ming Dynasty was characterized by the complete centralization of all power in the hands of the emperor and a powerful secret police. It adopted policies of reducing contacts between Chinese and foreigners and of stopping private ventures overseas because this would lead to centrifugal coastal centers of power. As the dominant recruitment subject, the Ming introduced a special type of essay which was rigid in form and elegant in style, but indoctrinating in substance and hostile to innovation in effect." (5)

Under the Ming Dynasty, the merchant class lost much of its social status as well as its professional autonomy. The imperial regime re-instituted the official orthodoxy of Confucianism and emphasized the ideal Confucian economy: self-sufficient villages with minimal commercial networks or specialization of labor. The Confucian Ming Dynasty was hostile to mechanical

---

166 Refer to Huang (1974, 1981) for an in depth study of the stagnation and decline of Ming Dynasty. Wesson (1978, p. 198) comments: "...... Commerce brings hand and brain together, as noted by Needham (1953), who saw the weakness of the merchant class in imperial China as the chief cause of the inhibition of science......"
167 Refer to Filesi (1972, p. 69) and Jones (1981, p. 205).
168 Refer to Eberhard (1960, p. 342, note 250), Filesi (1972, pp. 32-3, 69, 71) and Jones (1990, pp. 10-11).

contrivances, demolishing, for example, the astronomical clock built in 1090 AD. The Jesuit Matteo Ricci found little indication in 1600 AD to show that mechanical clocks had ever existed in China.[169] The Chinese turned their minds away from technology and industry, with the Ming Dynasty refocusing Chinese energies back into agrarianism for reasons of internal security. These policies led the Chinese empire on a path of transformation from a highly advanced economy (developed during the Song Dynasty), into a conglomeration of self-sufficient villages.

The government of the Ming Dynasty was one of minimal governance: the ideal form of rule advocated by the Confucian literati. The service state was dismantled, and in its place was a central government that performed mostly religious or ceremonial functions and provided moral leadership. Effectively, state officials served more as hierophants than administrators. In fact, the Ming and Qing Dynasty governments bore a greater resemblance to the Catholic Church than to the governments of European national states.[170] It was astonishing that a continental-sized empire was governed mainly through the force of culture and especially moral teachings.

Practically, the state of the Ming and Qing Dynasties existed solely to serve itself and keep itself in power. Economic productivity, as well as scientific and technological advancement, suffered greatly.[171] Apart from the need to serve as a defense against inner Asia, the emperor kept an army primarily to safeguard his own interests. This included the defense of the Grand Canal, which solely served him by enabling his important assigned grain tribute to reach the court of Beijing. The emperor had a lesser need for a military budget than European monarchs, and that allowed him to be less dependent on the merchant class to finance his ambitions. The merchants consequently received minimal concessions.

The total budget of the central government was not large. At the end of the nineteenth century, it was only one or two per cent of the national income.[172] The investment on infrastructure was trivial, at around 0.03-0.06 percent of the national income.[173] The Chinese paid about 24 percent of their national income mostly to the local elite (and to the central elite), who made up about two percent of their number, in return for essentially only two services: defense and the coordination of irrigation and flood control. No other important services, such as civil policing, were provided. The Ming and Qing governments also failed

---

169 Refer to Gimpel (1977, p. 15?)
170 Refer to Stover and Stover (1976, pp. 135, 186).
171 Refer to Wesson (1967, 1978), Huang (1974, 1981) and Jones (1981, 1988, 1990).
172 Refer to Perkins (1967, p. 487).
173 Refer to Stover and Stover (1976, p. 113).

to standardize weights and measures. In contrast, Lord Shang of the Kingdom of Qin during the Warring States Era of China had standardized weights and measures, and so had Napoleon of France of early nineteenth century Europe.

In sum, the Chinese empire of the Ming and Qing Dynasties was essentially an Asian revenue pump and nothing more, with no incentive to provide public goods and services to the society and economy for greater social welfare or for higher productivity.[174] The bureaucratic infrastructure of the state was kept small and cheap and was incapable of controlling the everyday life of the peasants or town dwellers. Although peasants and town dwellers were haphazardly exploited and neglected, they were not systemically repressed.[175] On the inefficiency and slackness of the Chinese gunpowder empire, Jones (1990) comments that:

"The survival of a lethargic government was probably assured by the relative absence of outside pressure... Notoriously, government did not provide much in the way of infrastructure or services... Government did not standardize weights and measures. Government did not provide commercial law or police... Government seems actually to have been withdrawing from participation in the economy in Qing times, preferring (for example) to let guilds carry out what we would consider, even what had once been considered in China, bureaucratic or legal functions. Certainly government did not mint enough money: token moneys had to substitute... Plenty of trade went on but there was a tendency for the market to work within the confines of personal acquaintance or the guarantees of native-place associations... We may find that the Qing economy, impressively expansible though it proved, failed to move from *extensive* to *intensive* growth because its political structure did not establish a legal basis for sufficient new economy activity outside agriculture... After the Song, China may not have recaptured, indeed may have moved further from, the successful organization of a compact, centralized state." (17-21)

The situation in Japan was quite different. Western gunpowder weaponry was first brought to Japan through Portuguese, Spanish and Dutch traders. With the introduction of gunpowder weaponry, there were greater economies of scale in warfare in Japan. Japan subsequently entered its Era of Warring States. The size of infantry legions maintained by the warlords greatly increased and the intensity of warfare prompted tactical innovations. Japan actually became

---

174 Refer to Jones (1981, pp. 206-209).
175 Refer to Moore (1967, p. 173), Jones (1981, p. 207) and Stavrianos (1982, pp. 287-289).

a forerunner in the gunpowder military revolution. Oda Nobunaga invented the use of volley fire, predating its use by the Dutch by some two to three decades.[176] Oda Nobunaga died before he could truly unite Japan. His chief lieutenant Toyotomi Hideyoshi achieved a short-lived unification of Japan. After the death of Hideyoshi, following a brief civil war, unification was re-established by the associate and archrival of Hideyoshi, Tokugawa Ieyasu.

In early modern Japan, the Kanto region – the plain centered on present-day Tokyo, was the largest core area. It commanded a significant share of Japan's population with economic resources that were larger than any other core area. On top of that, the Kanto region was strategically located. It seated in the center of Honshu Island, the largest and most important of the Japanese archipelago. This strategic location gave its master, the Tokugawa Shogunate, the benefit of easily controlling all other major regions of Japan. Furthermore, the Kanto region was also moderately distanced from the cluster of population and economic centers surrounding the Seto Inland Sea (cities such as present-day Kyoto, Osaka, Kobe and Hiroshima). The detached location allowed the Tokugawa Shogunate the benefit to avoid embroilment in the intricate balance of power games among the power centers crowded in southwestern Japan. The Tokugawa Shogunate therefore enjoyed moderate military and economic advantages versus other contestants. With the larger mass factor created by the gunpowder military revolution, the Tokugawa Shogunate managed to achieve a higher concentration of resources and capability ended the highly fragmented and pluralistic political order of medieval Japan.

Japan after the gunpowder revolution had a moderately high concentration of resources and capability. About half of Japan's population was under the control of the Tokugawa Shogunate or its allies. However, there remained many highly independent feudal lords, though they were nominally under the authority of the Tokugawa Shogunate. These more independent lords could be found especially in the southwestern part of the country, centering on the Seto Inland Sea. The mass factor was moderately large under gunpowder military technology. Yet, given Japan's fragmented and multiple core areas geography, warfare was not as decisive as in China or the Middle East given the same military technology. The moderately high concentration of capability and moderate size of the mass factor resulted in a relativist concern that was moderately low yet considerably higher than that of China under the Yuan, Ming and Qing Dynasties. Furthermore, the moderately high concentration of capability together with the moderate size mass factor generated a significantly risk-averse power-induced risk attitude which caused substantial distortions in economic decisions.

---

176 Refer to Parker (1996).

Together, these two mechanisms caused imperial complacency and conservatism and a declining performance of the Japanese economy and society.

Under the Tokugawa Shogunate (1603-1868 AD), social and economic stability were maintained by strict segregation of farming and trade and a ban on private investment. The officialdom discouraged any contact between different parts of the country that did not use the closely controlled Five Highways. Foreign trade and contacts were also discouraged.[177] Under Tokugawa rule, the expanding Japanese overseas undertakings of the previous era of political fragmentation were outlawed. Like Ming and Manchurian era China, Japan had chosen isolation, and like China, this only changed when Japan was incorporated into the Europe-centered modern world state system. In Japan's case, this change happened with Commodore Perry's naval expedition to open Japan's doors by force, in 1853 AD.

The Tokugawa government's policy of international isolation, active discouragement of commercial investment, and restrictions on communications and trade between domains combined to restrain the development of domestic markets. Neo-Confucianism formed an important part of Tokugawa official doctrine, which stressed morals, education and hierarchy. The regime adopted an anti-merchant policy and fixed the social status of groups. Merchants, who were considered to be parasites living off the labors of others, found themselves at the bottom of this hierarchy. The shogunate repeatedly made attempts to stem the tide of urban expansion and merchant success by repatriating peasants, confiscating property, and exiling wealthy merchants. Furthermore, the shogunate diligently issued legal proscriptions against townspeople acquiring obvious signs of wealth in an attempt to force people (particularly those residing in Edo) to conform to the Confucian standards of their lowly status.

Yet, despite repeated efforts to enforce a strict social hierarchy, the position of the Tokugawa government steadily declined. Public finance was in constant decline, and the steady worsening of the public financial situation led to higher taxes and peasant riots. The merchant class grew more powerful, which led to increased breakdown in the social hierarchy as samurais became financially dependent on the merchants. Corruption, incompetence and decline of morals were widespread within the Tokugawa government.

An important reason for the decline of the Tokugawa government was that the political structure of Tokugawa Japan was highly decentralized and consequently the shogunate failed to truly control the regional economies and inter-regional economic flows. During the Tokugawa era, regional feudal lords still retained much political, fiscal and economic autonomy. There was,

---

177 Refer to Haywood (1997, pp. 230-231).

therefore, a certain level of political competition and a significant level of mercantilist competition between the different regions.[178] During the extended
peace of the Tokugawa Shogunate, there were many improvements in farming
technology and practice. Cash was diffused throughout the Japanese economy;
inter-regional commerce led to regional specialization; the development of
specialized crops further contributed to considerable growth of the domestic
economy. All these advances happened despite central government policies
and weakened the central government's position versus the regional lords and
the merchant class.[179] Thus, although the political system itself weakened and
stagnated through the years of the shogunate, in other ways Japan saw somewhat surprising advances.

The next case study is the Indian subcontinent. In South Asia, the Indo-Gangetic Plain of northern India was the largest core area. This core area had a
significant share of South Asia's population and economic resources. There was
therefore a substantially asymmetric relative combined military and economic
efficiency between the contestant controlling this dominant core area and the
many other contestants holding other smaller core areas. With the larger mass
factor created by the gunpowder military revolution, there emerged in South
Asia a higher concentration of resources and capability. The highly fragmented
and pluralistic political order of medieval India was brought to an end.

It was the Delhi Sultanate that first ended the medieval fragmented order.
The Delhi Sultanate used gunpowder weaponry in the latter part of its military campaigns and that destructive power enabled the sultanate to unite the
northern Indian Indo-Gangetic plain whilst also briefly extending its rule to the
Deccan highland.[180] However, it was the Mughal Empire that fully capitalized
on the destructive power of gunpowder and cannons to establish a lasting continental-sized empire in South Asia. The reign of the Mughal Empire was officially from 1526-1858 AD, although it effectively ruled over most of South Asia
from 1572-1707 AD.

The Mughal Empire reached its peak in the 17th century. Although its fron-
tiers were occasionally pushed back, its power was largely unchallenged, and
the Mughal Empire had a much higher degree of administrative penetration
than that of the Delhi Sultanate.[181] Highly confident of the imperial power and
stability established by Akbar, Mughal statesmen subsequently did not see any
need for reform. Instead, the wealth of the empire was squandered in displays
of splendors and luxury, making the Mughal courts among the most dazzling

---

178 Refer to Distelrath (2004, pp. 107-108) in Bernholz and Vaubel (2004)
179 Refer to Jansen (2002).
180 Refer to Kulke and Rothermund (1998, pp. 158-169) and Khan (2004).
181 Refer to Kulke and Rothermund (1998, pp. 169,184-223).

of all the oriental courts. As its leaders played with their riches and neglected governance, the empire consequently declined.[182]

The deterioration of the Mughal Empire was not surprising. Given the large mass factor of the gunpowder military technology and the very high concentration of capability in the Indian system, relativist concern was very low. Furthermore, the very high concentration of capability and the large mass factor generated an extremely risk-averse power-induced risk attitude which caused severe distortions in economic decisions and bred extreme imperial conservatism. Together, the two mechanisms of imperial complacency and conservatism led to a very huge change in the performance of South Asian economy and society. The decline in performance from the highly fragmented and pluralistic medieval international political order of medieval India was very drastic.

During the long reign of the Mughal Empire it failed to provide any major services to benefit the economy.[183] The total tax revenue of the state was from fifteen to eighteen per cent of national income, largely collected as a land tax.[184] The bulk of this extracted revenue was consumed by the ruling elite, and almost none of it was spent on the provision of infrastructure or other public goods and services for social welfare or higher productivity. Furthermore, taxes were collected regardless of the state of the harvest. This callous practice often left the peasants destitute in the face of frequent natural disasters and state ineptitude and indifference. Notably, there was a horrific run of famines between 1540 and 1670 AD, even though the empire was at peace.[185]

The Mughal Empire had no written legal code. Neither was there any institutional design to harmonize contradictory orders issued by different rulers. Consequently, many orders were issued by one ruler only to be countermanded by the next. The efficiency of the empire therefore depended mainly on the character and ability of the Mughal emperors and more often than not, de facto power slid into the hands of serving officials. With no constitutional checks on the rapaciousness of these officials, the Mughal Empire became a typical predatory state that later proved itself incapable of standing up to the encroaching East India Company. Mughal India was colonized by Britain in the early eighteenth century and, like Japan and China about a century later, incorporated into the Europe-centered modern world state system.[186]

---

182 Refer to Berinstain (1998, p. 73).
183 Refer to Wesson (1967, p. 296) and Rothermund (2004, pp. 142-146) in Bernholz and Vaubel (2004).
184 Refer to Maddison (1971) and Jones (1981, p. 198).
185 Refer to Jones (1981, p. 187).
186 Subrahmanyam (1989) and Nadarajah (1992) argue that the post Mughal Indian states system were on the verge of creating a modern high fiscal-capacity state in the

The next case study is on the eastern Mediterranean and Middle Eastern region, an area that was dominated by the Byzantine Empire and then the Ottoman Empire during the medieval and early modern era. In southeastern Europe, the eastern Mediterranean basin, and the adjacent parts of Middle East, the most important core area was the region comprising Greece and Anatolia, centered on Constantinople or Byzantium. This core region was the power basis of the Byzantine Empire, an imperial tradition that lasted well over a millennium. Besides the combined resources of Greece and Anatolia, which were quite substantial in the southeastern Europe, eastern Mediterranean and Middle Eastern regions, the power holding Greece and Anatolia also controlled the highly lucrative maritime trade route between the Black Sea, the Aegean Sea and the Mediterranean Sea. Constantinople itself controlled the choke point of the Straits of Bosporus. In contrast, other major core areas in the region, for instance, the Nile River Valley or Mesopotamia, had comparable agrarian resources but lacked the control over the highly lucrative maritime trade route. The joint region of Greece and Anatolia therefore conferred upon its master a highly favorable and asymmetric relative combined military and economic efficiency vis-a-vis his rivals. Despite these geographical advantages, by the late medieval era, given the small mass factor and the cavalry dominance of the battlefield, political fragmentation was the norm and the Byzantine imperial power had dwindled to almost nothing and the empire barely survived.

The gunpowder military revolution then raised the economies of scale in warfare. Given the substantially asymmetric relative combined military and economic efficiency between the contestant controlling the Greece and Anatolia region and the many other contestants holding other core areas, the larger mass factor created by the gunpowder military revolution generated an extremely high concentration of resources and capability. The Ottoman Empire very successfully exploited the destructive capacity of gunpowder in its imperial pursuits, being one of the first to employ muskets and cannons. The walls of Constantinople, almost impenetrable for over a thousand years, were brought down within hours by cannon fire. The Ottoman Empire took over Constantinople, renamed it Istanbul and made it the capital of the empire. A new imperial power was ushered into the Middle Eastern and eastern Mediterranean region. The Ottoman Empire was the Islamic successor of the Byzantine Empire and soon proceeded to acquire the former extensive realm of the Byzantine Empire and beyond.

The Ottoman Empire was a very stable and uncontestable imperial order, reigning over most parts of the Middle Eastern region from 1280-1922 AD. The long

---

case of Mysore at the dawn of European colonial expansion. The establishment of the British Indian Empire disrupted this process of war making and state making before it could give rise to indigenous modern states in India.

reign of the Ottoman Empire was roughly comparable to the combined duration of the Mongolian Yuan, Chinese Ming and Manchurian Qing Dynasties in China. At the peak of its power, the Ottoman Empire controlled the Balkan Peninsula, Hungary, the territories around the Black Sea, northern Africa (excluding Morocco), the Red Sea region, the Levant, Cyprus, Mesopotamia and the southern coast of the Persian Gulf, besides its core area of Greece and present-day Turkey. That was an empire larger than the Byzantine Empire at its peak under Emperor Justinian, and comparable to the Roman Empire at the height of its power. The Ottoman Empire achieved a very high concentration of resources and capability indeed.

Given the very high concentration of resources and capability in the Ottoman system and the large mass factor of the gunpowder military technology, the Ottoman Empire had a very strong risk-averse power-induced risk attitude and a very low concern for marginal relative capability. The highly risk-averse power-induced risk attitude caused the Ottoman authorities to be very conservative and the very low relativist concern made the Ottoman Empire extremely complacent. The Ottoman Empire was lethargic. It did not advance economically nor did it produce any significant cultural achievements.[187] The dismal performance of the Ottoman Empire was unsurprising. There was a stark decline in performance from the highly fragmented and pluralistic medieval international political order of the medieval eastern Mediterranean and Middle Eastern region under the Abbasid Caliphate and Seljuk Turks. Unlike European sovereigns who secured their positions by offering the service of justice, the Ottoman Empire offered very few overhead services to the proper functioning of either the economy or society. The Ottoman Empire, though of immense territorial size, had a small state without much capacity to penetrate and mobilize the society for its purposes.[188]

The Ottoman Empire failed to develop or maintain mercantilist policies, measures that would have been necessary to halt the economic penetration of European producers and to sustain state power in the international arena. For instance, embargoes on the export of strategic goods to Europe were not enforced as state bureaucrats were simply too inefficient and corrupt to enforce policies in the interest of the state. Consequently, the Ottoman regime even failed to procure essential war materials and food required by its own armies. From the eighteenth century onwards, the weakened Ottoman Empire was always on the defense on its European front, under incessant assault from the great European powers of France, England, Russia and Austria.[189] The lethargic Ottoman Empire was eventually referred to as the 'Sick man of Europe'.

---

187 Refer to Jones (1981, p. 189) and Kuran (2004, pp. 153-156) in Bernholz and Vaubel (2004).
188 Refer to Jones (1981, p. 180) and Goldschmidt (2002, Ch. 9).
189 Refer to Armstrong (2000, pp. 97-117).

In Central Asia and the Middle East, the most important area was the Iranian Plateau. This core region had earlier formed the power basis of the Achaemenid Empire, the Parthian Empire and the Sassanian Empire. The tradition of imperial authority and unity was well grounded in this region: this was a political tradition with a history of over a thousand years before the Islamic conquest. Compared with the surrounding regions, this core area had a share of resources which was quite substantial and dominant. After the Islamic conquests, given the small mass factor of medieval military technology, political fragmentation was the norm in this region.

As in the other major Eurasian cultures, the gunpowder military revolution enlarged the mass factor in the Persian system. Given the substantially asymmetric relative combined military and economic efficiency between the contestant controlling the Iranian Plateau and the other contestants, the larger mass factor generated a higher concentration of resources and capability. The Mongolian Ilkhanate was the very first empire in this region to exploit the destructive capacity of gunpowder in empire-building, and it started a new tradition of imperial unity and authority in the Persian system.

The Ilkhanate set up by Prince Hulegu reigned from 1256-1335 AD and was followed by a succession of empires: the Khanate of Timerlane (ca 1370-1507 AD); the indigenous Safavid Empire (1501-1722 AD); the Afshar Dynasty (1729-1747 AD); the Zand Dynasty (1750-1794 AD) and finally the Qajar Dynasty (1795-1925 AD). The victory of Shah Ismail of Safavid Persia at Mervover Muhammad Shaybani of the Shaybanid Khanate of Uzbek on December 2nd, 1510 AD, was considered a turning point in the struggle between nomadic and sedentary civilized societies. The firepower of gunpowder weaponry defeated the horse power of cavalry and the sedentary society began to gain the upper hand against the nomads.[190] With the aid of gunpowder weaponry, civilization started to establish its dominance over the steppe again.

The Persian system after the gunpowder revolution had a very high concentration of resources and capability. Furthermore, the mass factor was very large, given the destructive power of gunpowder military technology. Consequently, Persian relativist concern was very low. The series of gunpowder empires starting with Ilkhanate were made more uncontestable by the larger mass factor. Despite several dynastic changes, the imperial order swiftly and successfully reconstituted itself time and again. Furthermore, given the very high concentration of capability and the very large mass factor, the power-induced risk attitude was highly risk averse, which caused severe distortions in economic decisions. Together, the very low relativist concern and very risk-averse attitude

---

190 Refer to Grousset (1970).

resulted in imperial complacency and conservatism in the Persian system and caused a dismal performance in the Persian economy and society.

Imperial complacency and conservatism plagued Safavid Persia, especially after the decline of the Ottoman threat. Corruption and decadence were omnipresent. Almost unbelievably, the maintenance of the huge imperial harem swallowed up close to half of the state revenue. Once the external environment was quite peaceful, even the central standing army was not properly maintained. Funds were diverted for the consumption of the harem. The army largely existed only on paper, was good only for parade, and certainly had no capacity for war.

In many key regional posts, appointees from the central government who bought their positions replaced local tribal chieftains. The local tribal chieftains, who had been highly autonomous, were efficient in military affairs and once the pillar of the military power of the empire. These replacements strengthened the position of the central government but weakened the defense capacity of the empire, and led to excessive extraction from the local economy. Many regions suffered greatly and depopulated. Great concentration of wealth at the court led to declines in trade and revenues. The royal treasury was a bottomless pit: resources were extracted from the economy, without the provisions of public goods or services in return. Sale of offices and tax farming resulted in corruption, incompetence and excessive extraction; justice was neglected and public safety was not maintained. Highway robberies were common, famines broke out even during peacetime, and monetary stability failed to be maintained. Debasement of coins happened frequently and fueled inflation; a shortage in the money supply caused the economy to be under-monetarized and trade suffered.

Power ceased to reside with the weak and ineffectual Safavid rulers, and instead passed inadvertently to the imperial harem and the Shiite religious leaders. The practice of appointing princes as provincial governors was abandoned for fear of formation of political factions around them and the threats posed to the emperor. Scions were brought up in the harem, totally ignorant of the outside world. Incompetence in statecraft or military science and overindulgences in pursuits of sensual pleasures and neglect of duty characterized Safavid rulers. The Queen Mother and forceful wives and concubines of the emperor, as well as eunuchs, began to rule. Imperial harem factional politics drove the history of the dynasty while outside the harem the persecution of religious minorities by the Shia ulama tore the social fabric. All of the above factors cut the Safavid Dynasty off from its supporters.[191]

Persia remained in permanent decline from the mid 17th century onwards. As in the Ottoman Empire, there were occasional efforts made to reform or

---

[191] Refer to Foran (1992) and D'souza (2002).

modernize the state, the economy and the army in Persia. However, sustained efforts remained elusive. When threats were not as severe or the external environment became peaceful, the empire returned to its lethargy and complacency and conservatism regained its supremacy in politics. Throughout the long period of decline, the small state apparatus remained as a revenue pump for the imperial court and nothing more.

The Safavid Empire was replaced by the Afshar Dynasty of Nadir Shah. Iranian military power had a strong resurgence under the charismatic leadership of Nadir Shah, a military genius described by some historians to be Napoleon of Persia or the Second Alexander. The Afshar Dynasty briefly controlled Afghanistan, western and southern Pakistan, Iraq, the southern coast of the Persian Gulf, Georgia, Caucasus, and the southern part of Central Asia to the shore of the Aral Sea.[192] Nadir Shah also invaded Mughal India, raided Delhi, and carried off bounty and booty, including the famous Peacock Throne. At this time Peter the Great of Russia probed the Caucasus area, but Nadir Shah repulsed the Russian incursions. However, once the charismatic leadership of Nadir Shah left the scene when he was assassinated in 1747 AD, his empire quickly disintegrated and the long decline of Persia resumed and continued. When Russia under Catherine the Great renewed expansionist efforts into the Caucasus, Georgia and Central Asia, Persia failed to resist successfully. By then, Persia was too weak and Russia had become too powerful, and Persia, or Iran, became an appendix to the western-centered modern global state system at the end of the Zand Dynasty.

4. Eurasia

After the establishments of the late medieval and early modern gunpowder empires, the major non-European cultures turned isolationist economically, due to the lack of external political-military competition and an imperial attitude of conservatism and risk-averseness. Consequently, the Silk Road had stopped serving as a major trade route by around 1400 AD. The decline of trade among the major oriental cultures was compensated for by the increased overseas trading activities of western nations.

Almost contemporary with the gunpowder military revolution were a series of maritime transportation revolutions and the great geographical discoveries of the new world and new shipping routes, all of which knitted the different

---

192 Refer to Haywood (1997, pp. 214-215).

and distant cultures of the whole globe into one large community and a global state system. Western powers reached the shores of the major oriental cultures soon after the great geographical discovery of the new world and new trade routes: first came the Portuguese, Spanish and Dutch, and they were soon followed by the French and English. But they were not in a position to challenge the gigantic oriental gunpowder empires and their presence was not viewed as a threat. With the exception of India, which was incrementally encroached upon and conquered by Britain when the Mughal Empire was dissolving and India was in the depths of civil war, no major oriental Eurasian culture was overrun by another Eurasian power (including the Western great powers) before the industrial revolution. The gravity of political-military competition on a mega Eurasian or global scale still lay with the interactions among the major Eurasian cultures.

The pan-Eurasian international political system worked more forcefully with the coming of the gunpowder era. In the 13th century, there were attempts at a Franco-Mongol alliance with exchange of ambassadors and even military collaboration in the Holy Land. The targets of the alliance were the Middle Eastern Islamic forces, especially the Egyptian Mamluk Dynasty. After the Mongol Empire ended, the political and military interactions among major cultures subsided, since the mobility of the Mongolian tribesmen and other nomads had been taken out of the equation. Part of the reason was that as the gunpowder military revolution proceeded, it ultimately curtailed the power of the nomads and their cavalry, as infantry equipped with gunpowder weaponry ultimately regained battlefield advantage and warfare became less mobile.

The logistical constraint of pre-industrial technology restricted political-military interactions among the major oriental empires on the Eurasian landmass. The heavy weaponry of early gunpowder technology and the massive infantry legions faced considerable mobility problems, especially when traveling overland. Consequently, political-military interactions among the major Eurasian cultures over long distances during the pre-industrial gunpowder era were hardly more substantial that those during the medieval era. In the sea, on the other hand, there were significant advances in maritime technology during the late medieval and early modern era.[193] Political-military competitions among the major cultures on and across the sea were more significant than those of the medieval era, though not yet of a frequency and intensity substantial enough to affect the developmental performance of the major Eurasian cultures.

The Mongolian Empire, not satisfied with its imperial pursuits overland, had sent military expeditions across the sea. However, given the Mongolian

---

193 Refer to Cipolla (1966).

unfamiliarity with maritime affairs, the Mongolian invasions of Japan in 1274 AD and 1281 AD ended disastrously. So did the Mongol invasion of Java in 1293 AD. Mongol invasions of Vietnam in mid and late thirteenth century through land and sea failed too. After the Mongol attempts, Japan invaded Korea twice under the Hideyoshi Shogunate: Ming Dynasty China intervened and Japan was defeated by the Chinese-Korean Alliance. The failures of these overseas military campaigns testified to the limit of naval power projection during the pre-industrial gunpowder era. Outside the Eurasian world, early and pre-industrial European projection of powers and conquests overseas were victories against societies of the Americas, Africa and Oceania, which were militarily much more backward—the Europeans of that time had little success against the other more advanced and established Eurasian major cultures.[194]

On the Eurasian land mass, the constraints imposed by pre-industrial logistical capacity restricted political-military interactions between China and the other major empires. Unlike the medieval era, when Tang Dynasty China had set up military outposts in India, after the gunpowder military revolution, China under the Yuan, Ming and Qing Dynasties had no substantial political-military interactions with India. During the medieval era, Tang Dynasty China competed with the Abbasid Caliphate over Central Asia. After the gunpowder military revolution, there was no political-military competition between China and Persia. During his last years, Timerlane contemplated attempting to conquer China but died before any action was undertaken.

There were more political-military interactions between the Ottoman Empire and the Persian Empire, and between the Persian Empire and the Mughal Empire. However, the Islamic Ilkhanate, Timurid Empire and the Safavid Empire and later successors, were steppe-highland, desert-oasis powers with cavalry still playing a significant role. In contrast, the Ottoman Empire, like its predecessor the Byzantine Empire, was a Mediterranean power. Besides having a powerful navy, the famous slave-elite troops of the Ottoman Empire, the Janissaries, was infantry. On the other hand, the Sultanate of Delhi and the Mughal Empire ruled over tropical and subtropical forests and monsoon river plains which necessitated a difference in war craft. Nadir Shah of Persia invaded India but retreated after raiding Delhi. Differences in geography and mode of warfare as well as great distance and logistical constraints mitigated the intensity of conflict between the three major Islamic empires: the political-military interactions among the major Islamic gunpowder empires therefore failed to significantly affect their developmental performance.

---

194 Refer to Parker (1996).

The Ottoman and Safavid rivalry was the most important of the political-military interactions among the major oriental gunpowder empires. Shiite and Sunni animosity led to clashes between the Ottoman and Safavid Empires during the sixteenth century, specifically over the control of Mesopotamia. However, the core areas of both empires remained unthreatened. This was almost a repetition of the rivalry between the Roman and Parthian Empires and between the Byzantine and Sassanid Empires. With the subsiding of religious ferment within the Safavid Empire, the Treaty of Qasr-e Shirin (1639 AD, also called the Treaty of Zuhab) established a permanent peace between the Ottoman and the Safavid Empires. Furthermore, in the sixteenth century, extensive clashes with the Uzbeks and other powers in Central Asia caused the Safavid Empire to look to the north more often than to the west. The advance of the Russian Empire further reinforced this northern orientation of the Safavid Empire: Safavid Persia fought a prolonged war with the Russians during the 1720s. On the other hand, with the advancement of the European powers, the Ottoman Empire permanently focused on securing its influence to the west and to the north, warring incessantly with the Russians and Austrians and the other European powers. Political-military competition between the Ottoman Empire and the Persian Empire therefore was tempered by geography and geopolitics and had little significant impact on the developmental performance of the two empires.

In sum, the gigantic oriental gunpowder empires were largely non-competitive towards each other. They did not threaten the survival of each other, though there might be non-military contacts or minor or brief military contests. It seemed that the equilibrium would persist. Yet, concomitant with the gunpowder military revolution were the advances in maritime technology, which included the use of the compass and the building of better sailing ships for ocean going.[195] The geopolitically separated worlds of the different major cultures were coming to an end. The only major Eurasian culture that escaped the fate of being a gunpowder empire with its associated imperial complacency and conservatism was Europe. Europe had a competitive state system and the energy generated by that system was going to disrupt the rough equilibrium among the major Eurasian cultures. Europe was going to knit the different isolated geopolitical systems of the whole world, Eurasian and non-Eurasian, into one global international political system.[196]

---

195 Refer to Cipolla (1966).
196 Refer to Buzan and Little (2000, Part IV) for the emergence of the modern global international system.

5. Europe

In contrast to the experience of the other major Eurasian cultures, the gunpowder military revolution failed to create a pan-European Empire and destroy the European pluralistic international order. Instead, the gunpowder military revolution increased the intensity of interstate political-military competition in Europe by furthering the rise of compact national states and eroding the two imperial powers in Europe, the Holy Roman Empire and the Papacy. The new technology altered the pluralistic but mildly competitive environment of medieval Europe and ushered in, not an imperial order but, a highly competitive state system.

Europe has a highly fragmented geography. Waters and mountains divide Europe into many easily defendable regions each with their moderate size core areas, with no single dominant core area retaining an overwhelming concentration of resources to wage a decisive war on a pan-European scale. The mass factor therefore was smaller in Europe than in the other major Eurasian cultures, given the same military technology. Furthermore, in Europe, partly due to the geographical and political fragmentation, there was no dominant linguistic or ethnic group. This cultural heterogeneity of Europe made unified political administration difficult and costly and further hindered any imperial pursuit. Last but not least, the existence of an offshore England as well as powers external to Europe (such as the Ottoman Empire, Russia and, later, America) also hindered the creation of a single imperial unity. Offshore England and the other external powers repeatedly ensured that all attempts to put Europe under one imperial design would be thwarted.[197] On this point, Montesquieu (1748) observes that:

"In Asia they have always had great empires; in Europe these could never subsist. Asia has larger plains; it is cut into much more extensive divisions by mountains and seas ... In Europe, the natural division forms many nations of a moderate extent, in which the ruling by laws is not incompatible with the maintenance of the state ... It is this which has formed a genius for liberty; that renders every part extremely difficult to the subdued and subjected by a foreign power." (Book 17, Ch. 6)

On the other hand, all societies in Europe had access to and were familiar with the Greco-Roman tradition, especially after the Renaissance. Classical science and knowledge, including laws and statecraft, were therefore equally available

---

197 Refer to Powell (1999) on the alignment and stability of the international system. Refer to Wesson (1978, p. 111), Kennedy (1987, Ch. 1) and Hui (2005)for a comparative study of how the early modern Europe state system survived while the ancient Chinese state system failed to.

to all. This cultural similarity, together with the absence of a dominant core area, meant that all the major European states were quite equal in relative combined military and economic efficiency. Consequently, despite the larger mass factor due to the gunpowder military revolution, Europe retained its state system and maintained a highly even distribution of resources and capability.

The gunpowder military revolution started to affect Europe in the fourteenth century: cannon were used by the Florentine army as early as 1326 AD.[198] Gunpowder weaponry was first effectively used in the early fifteenth century during the battles of the Hussite Wars (1419-1434 AD). As in the other major cultures, the gunpowder military revolution increased the economies of scale of warfare in Europe. In Europe, from the sixteenth century on, there were series of innovations in military technology that increased the size of armies and the scale of warfare brought about by the use of gunpowder weaponry and associated new tactics, referred to as the military revolution.[199] Armies grew larger, with infantry equipped with gunpowder weaponry playing a more important role.

The gunpowder military revolution transformed the European states, economies and societies. Especially from the sixteenth century onwards, growth in the size of the armed forces and military budget was faster than growth in the size of the population and economy. Between 1530 and 1710 both the total numbers of armed forces paid by the major European states and the total numbers involved in the major European battles increased tenfold.[200] Large standing armies and navies emerged in the 16th and 17th centuries due to the use of gunpowder weaponry. The use of cannons forced the replacement of medieval forms of fortification by the larger and more expensive fortifications called trace Italian.[201] European states competed with each other to field larger and larger armies and navies, limited only by the resource extraction ability of the state and the size of the economy. Increased scale in warfare increased the financial burden of the state. To meet the requirement of fielding larger armies and navies and extracting more resources for military purposes, the European states resorted to many measures to outperform each other economically, fiscally and financially.[202]

The larger mass factor caused compact, centralized national states to replace their many competitors.[203] Feudalism was abolished. Absolutism became the guiding principle for monarchs. The increase in the economies of scale in appli-

---

198 Refer to Haywood (1997, p. 168).
199 Refer to Roberts (1956), Parker (1976, 1996), Duffy (1980), McNeill (1982), Black (1991), Dudley (1991, 1992), Tilly (1992) and Keegan (1993).
200 Refer to Parker (1976).
201 Refer to Roberts (1956), Parker (1976, 1996), McNeill (1982), Black (1991), Dudley (1991) and Porter (1994).
202 Refer to North (1995, pp. 13-17; 1998, pp. 16-19).
203 Refer to Spruyt (1996).

cation of violence and coercion reduced state costs in suppressing local feudal resistance, and facilitated the enforcement of property rights enforcement in centralized nation-states.[204] Consequently, the race for a larger economy and revenue to support the larger military led to a larger state and greater state intervention in the economy and the rise of centralized absolutist national states.

Organized around a major economic core area, the national states of England, France and Spain defeated the many surrounding feudal principalities and expanded at their expense. Alliances of cities, such as the Hanseatic League, or loosely organized empires, such as the Holy Roman Empire or the religious transnational empire of the Roman Church, ultimately lost out to the centralized and compact national states which had greater fiscal extraction capacity and greater combined relative military and economic efficiency. The rise of the national states and the diminishing relevance of the two imperial powers led to a more even distribution of capability and a pluralistic international order in Europe. Consequently, conservatism receded. First it was the Reformation which challenged the traditional authority and wisdom of the Church. Then further cultural liberalizations arrived, such as the Enlightenment movement. The spirit of scientific enquiry as well as creativity in arts and humanity prevailed over the conservatism of traditions and religious doctrines.

With the medieval even distribution of capability preserved or even enhanced, the larger mass factor of the gunpowder military revolution increased the intensity of political-military competition in the European system. Defeats in wars became more costly. Weaker states or principalities were weeded out at a faster speed. Large centralized sovereign nation-states advanced at the expense of the motley landscape of feudalism and the number of principalities declined as the gunpowder military revolution proceeded. The size of political units grew as the feudal order retreated, because the smaller princely states and the city-states were too small to exploit the greater economies of scale afforded by the gunpowder military revolution. By the late fifteenth century and early sixteenth century, with the French and Spanish invasions of Italy, the mini-state system of Italy ceased to exist independently and was subsumed into the European state system. With a greater mass factor, political and military contests within the European state system became more frequent and decisive.[205]

The gunpowder military revolution failed to create a pan-European empire. Attempts to unite or dominate Europe were thwarted again and again. Charles V the Holy Roman Emperor of Habsburg failed; so did Louis XIV, Napoleon, Wilhelm II

---

204 Refer to Tilly (1975, 1992), Duffy (1980), Hechter (1980), Cohen, Brown, and Organski (1981), Best (1982), Black (1991), Anderson (1992), Downing (1992), Porter (1994) and Parker (1996).
205 Refer to McNeill (1982), Dudley (1992), Tilly (1992), Keegan (1993) and Parker (1996).

and Hitler. Instead, the more decisive new military technology locked the European sovereign nation-states in a seemingly perpetual and intense interstate rivalry, both on and off the battlefield. Consequently, a larger mass factor with a highly even distribution of capability generated higher relativist concern. Furthermore, there was no extreme power-induced risk attitude to distort decision making since the distribution of capability in the European international political system was highly even. Without the inhibition of imperial conservatism due to an extreme risk-averse power-induced risk attitude, the higher relativist concern generated by the intense European interstate power struggles ran free to propel forcefully the cultural and economic advancement of Europe.[206]

The absolutist states formulated forward-looking policies; statesmen, scholars and private individuals produced guidance for public policy and published exhortations to purposeful development such as "growth programmes" or virtual national plans. An example of the latter is Phillip von Hornick's *Austria Over All, If She Only Will* (1684).[207] Another good example is Friedrich List, who advocated the protection of infant industries to facilitate the industrialization of Germany as a counterweight to the economic dominance of England. The English measured themselves by their successful Dutch cousins. A seventeenth century tract on economic improvement was entitled, *How to Beat the Dutch without Fighting.*[208]

The European states' concern for fiscal strength and economic might led to the rise of many intellectual schools such as mercantilism, cameralism (which as the German counterpart of the French mercantilism), physiocracy and ultimately, classical economics. Amidst such a competitive international environment, mercantilism emerged as a school of thought and became popular, especially among the policy-making circles. Mercantilism proposed the subjection of all economic interests to the needs of the state or kingly power, and was especially concerned with the need to replenish and maintain the state war chests to finance the military.[209] France, under Louis XIV, applied mercantilism most thoroughly. The ideal model of mercantilist policy was France under Colbert, the Minister of Finance of France from 1665 to 1683 under Louis XIV. He improved the state of French manufacturing and brought the economy back from bankruptcy by raising tariffs, encouraging major public works projects, improving roads and canals, increasing the indirect taxes from which the privileged could not escape, establishing new industries, protecting inventors, inviting in workmen from abroad and prohibiting French workmen from migrating. He also founded the Academy of Sciences as well as other academies.

---

206 Refer to Jones (1981, pp. 113-119).
207 Refer to Wesson (1978, pp. 181-185), Jones (1981, p. 134) and Weiss and Hobson (1995).
208 Refer to Coleman (1961, p. 45) and Wesson (1978, p. 182).
209 Refer to Heckscher (1955), Wesson (1978, p. 136) and Webber and Wildavsky (1986).

French industrial output and economy grew considerably under Colbert and France became the dominant power in Europe.

The practice of mercantilism was not restricted to France. Most Western states embraced it to a certain degree, including Sweden, Courland, Denmark and Prussia. Colbert's policies also inspired those of the first treasury secretary of the United States, Hamilton.[210] Even the liberal and *laissez-faire* market-oriented Britain undertook enough state intervention in the economy for it to be labeled 'parliamentary Colbertism'.[211] Good examples were the Navigation Acts, which protected the British shipping industry. In fact, given the greater efficiency and fiscal capacity of the British state, Britain practiced mercantilism far better than any other European state, with the possible exception of France under Colbert.[212] This was perhaps not surprising, considering that London merchants held political power in the English parliament. This meant that commercial interests were able to press for the formation of courts to try and hence control backsliding officials. In contrast, corruption ran rampant on the European continent and greatly compromised state capacity. Britain became decidedly more protectionist in the era of William III, with expanded mercantilist policies to protect both home and colonial markets, in particular the infant linen and silk manufacturing industries.[213] Mercantilist policies helped to turn Britain into the dominant trader and military power globally.

An important cause for the rise of England was the early consolidation of the centralized national state. The centralized state of England aided the development of the capitalist economy through its early sponsorship of a legal framework friendly to the functioning of the market economy. Consequently, the English common law system developed precociously ahead of her European continental counterparts and with Anglo political and economic influence, spread to other parts of the world to become the most important functioning legal system today. On the other hand, the lack of or late development of a powerful centralized state in France, Italy, and elsewhere delayed and hampered development of capitalism on the European continent itself.[214]

Mercantilist policies helped national economies to modernize, with many of the policies explicitly designed to develop an industrial base. The mercantilist spirit caused the rise of the service state.[215] Rulers, whose pursuit of power and glory drove them to prepare for war, began to do so by actively improving

---

210 Chernow (2004, p. 170).
211 Refer to Weiss and Hobson (1995).
212 Refer to van Klaveren (1969), Brewer (1989) and Weiss and Hobson (1995).
213 Refer to Davis (1965) and Weiss and Hobson (1995).
214 Refer to Tigar and Levy (2000).
215 Refer to Bronfenbrenner (1964, p. 363), Landes (1969), van Klaveren (1969), Anderson (1975), Rostow (1975), Tilly (1975, p. 73) and Jones (1981, pp. 134-9).

the economic base critical for power and war. Mercantilist-absolutist regimes introduced new crops, notably the potato, and encouraged settlement of new lands. Mercantilist-absolutist states intervened to help societies to cope with natural disasters and epidemics, as well as supply public services such as fire fighting and the management of lighthouses. They also took measures to codify laws and they established standard weights and measures and unified the coinage. For instance, the French Civil Code of 1804, directed by Napoleon himself, was the most famous of all efforts to codify the laws. The genesis of the modern law codes however was probably the Prussian code, directed initially by Frederick the Great, which was in effect until 1900.[216] All these measures reduced transaction costs and boosted the economy. In contrast, this drive to trade and govern well for the sake of greater wealth and revenue was conspicuously absent among the leaders of the oriental gunpowder empires or the ruling elite of the states of noncompetitive state systems in Southeast Asia, America, Africa, Australia and Oceania. Wesson (1978) observes:

> "......While Asiatic emperors have been quick to despoil riches as soon as they become tempting, European kings have been able to do so only under some restraint and at considerable ultimate cost to their power......." (195)

The competitive international environment of Europe produced many diligent rulers, of whom Frederick the Great of Prussia was among the more famous. In contrast, the lethargic empires produced many rulers who had contributed next to nothing to their state or society. The Shen Zong Emperor of Ming China (1563-1620 AD), for instance, was absent totally from his court and had no involvement in the administration of his state for over thirty years.[217] Such contrasts give a good picture of the different effects that political military competition or the lack of it had on Europe and the other major Eurasian cultures.

Amongst European states, the constant jostling for power, position, and prestige in an insecure competitive state system generated an atmosphere propitious to improvement. Statesmen and public leaders were continually spurred by the need to improve at the price of their comfort, in marked contrast to the opulent courts of China or harems of Persia. Every advance of a rival was seen as a threat and a powerful goad. The entwined and competitive international system of Europe provided forceful prodding to keep European states and societies from complacency, preventing the rot of stagnation, decline and decay

---

216 Refer to Kempin (1990, p. 113).
217 Refer to Huang (1988).

that plagued the secure and entrenched gunpowder empires of the other major Eurasian cultures.[218]

In their efforts to outdo each other, the European states goaded each other into the great cultural, economic, political and social achievements that define the industrial and modern age.[219] The interstate rivalry caused the European states to try to outdo each other in almost all fields of human endeavor: overseas exploration, manufacturing, scientific enquiry, technological innovation, as well as improvements in law, public administration and the overall institution of the state. The effort of Cardinal Richelieu to create a great and powerful central government in France, the mercantilist policies of Colbert, the financial innovations of England such as the creation of Bank of England—all have their origin in the interstate great power rivalry of Europe.[220]

France was driven by its rivalry with England during the Napoleonic wars to promote manufacturing. In addition to the powerful stimulus of demand for industrial goods due to the war, another impact upon France's development was Napoleon's personal undertaking, in an effort to meet the British challenge, of the implementation of many policies and incentives to encourage inventions. Similarly, in her rivalry with Frederick the Great, Maria Theresa of Austria undertook many measures to improve the power of Austria relative to Prussia, including the modernization of laws and public administration. The best-known story of such leader-driven modernization is Peter the Great, who traveled through Western Europe, especially the Netherlands, to learn western secrets so as to strengthen Russia. His forceful westernization and modernization of Russia made it a great European power. The remark of Mokyr (1990) gives a good overall picture of Europe at the time:

> "......Western technological creativity rested on two foundations: a materialistic pragmatism based on the belief that the manipulation of nature in the service of economic welfare was acceptable, indeed commendable behavior, and the continuous competition between political units for political and economic hegemony......" (302)

The intense interstate rivalry spilt over to cultural and educational arenas.[221] Shattering the pretensions of Spain in the sixteenth century inspired

---

218 Refer to Gibbon (1932, Ch. 38, II) and Wesson (1978, p. 182).
219 Refer to Weiss and Hobson (1995) and Pollard (1998, pp. 223-237), Hartwell (1998, pp. 239-241)in Bernholz et al. (1998).
220 Refer to Rostow (1974), Kennedy (1987), North (1998, pp. 13-28) and de Vries (1998, pp. 209-221) in Bernholz et al. (1998).
221 Refer to Stomberg (1931), Roach (1969), Hale (1971) and Wesson (1978, p. 182).

Elizabethan poetry and drama, much like how it was when the Greeks had been uplifted by victory over Persia millennia before. At the beginning of the fifteenth century, the Florentines were uplifted by their successful defense of liberties, observed Baron (1955). But defeat as well as success could inspire change: the debacle of defeat in the Franco-Prussian War of 1871 AD led to increased self-examination efforts and reforms by the French, which also ushered in an efflorescence of art in France.[222]

More importantly, European states were quick to realize the importance of scientific and technological prowess and the education level of the population in determining the power of nations. Aiding this process of renaissance in learning and greater emphasis on knowledge was the use of the printing press, which facilitated the spread of information. In 1455 AD Johannes Gutenberg developed the movable type press, leading to the first printed books.[223] As education became available to the masses, vernacular languages replaced Latin and further aided the rise of sovereign nation-states, especially in post Reformation protestant Western and Northern Europe. Furthermore, competition between European states led to the revival of Roman law, which facilitated centralized administration. The nation-states in Europe therefore developed the characteristics of being well-organized, having strong governments, and possessing distinctive languages and strong national identities. In sharp contrast, the invention of the printing press, which made the spread of knowledge easier, led to a greater effort of censorship and not scientific or educational renaissance in the gunpowder empires of the Middle East, South Asia, and East Asia. Only until 1720 was the printing press used in the Arabian world. Differences in the international environment elicited totally different responses to new technology.[224]

In Europe, national education policy became a political necessity from the eighteenth century on, with education becoming more practical and science-orientated.[225] The Swedish Academy of Sciences, founded in 1739, for example, emulated English, French, and German models with the explicit aim of promoting national power, particularly economic power. Prussia (and Germany) and France (especially from the time of Napoleon) were big players in this race for scientific improvement.[226] After its defeat in the Franco-Prussian War, which ended French hegemony in continental Europe and resulted in the creation of a

---

222 Refer to Kavolis (1972), Tilly (1975, 1992), Duffy (1980), Jones (1981, 1888), Kennedy (1987), Rasler and Thompson (1989), Porter (1994), Weiss and Hobson (1995) and Parker (1996).
223 Refer to Haywood (1997, p. 145).
224 Refer to Stomberg (1931) and Wesson (1978).
225 Refer to Roach (1969) and Wesson (1978).
226 Refer to Wesson (1978, pp. 119, 141-152).

unified Germany, France reformed its education system.[227] Germany in turn responded by spending more on science and education. The nineteenth century German research universities played an important role in the economic ascendancy of Germany in the late nineteenth century. They were so successful that they were emulated by many of the most prestigious universities in the United States and elsewhere.

The same great power rivalry prompted the European states to undertake geographical explorations. Henry the Navigator of Portugal sponsored overseas exploration and he was not alone among the European ruling elite to do so.[228] Two of the major objectives of Henry the Navigator were to surpass the other European sea powers and to outflank the Islamic world which not only posed a military threat but also monopolized the lucrative oriental trade. Through colonial activities and overseas trade, the European state system was expanded to every corner of the globe to become the modern global state system.[229] W. W. Rostow (1960, p. 6, 32) argues that industrialization was "given dynamism by the lateral expansion of world markets, (i.e. the new overseas territories) and the international competition for them" and that "the meaning and impact of this lateral innovation was heightened and given a peculiar turn because it occurred in a system of inherently competitive nation states."[230] This expansionist energy of Europe contrasted sharply with the lethargy and isolationist orientation of the gunpowder empires in the Middle East, India, China and Japan. The Ming court of China, for instance, forbade further overseas expedition after the maritime undertakings of Admiral Zheng He. Taiwan was not colonized under the Ming Dynasty, even though it had been discovered for a long time and was within easily reached distance off the shore of China.

The destination of this metamorphosis was modern warfare, modern state and modern economy. Ultimately, Europe evolved as a group of states governing industrialized economies, with great industrial and technological power and mass participation in politics. These mass participation states organized and maintained massive armed citizen forces facing each other in long periods of militarized peace punctuated by large-scale conflicts. These changes affected naval warfare and subsequently aerial warfare too, which became even more expensive and more technology-intense, and had even greater logistical and financial demands on the state and the economy than land warfare.

---

227 Refer to Wesson (1978).
228 Refer to Findlay and O'Rourke (2007) on how military conflicts and geopolitics affected international trade.
229 Refer to Bernholz et al. (1998, p. 10).
230 Refer to Findlay and O'Rourke (2007) on how control of long distance trade, overseas market and raw materials aided the Industrial Revolution in Britain.

Intense international competition for power and survival instilled a fierce nationalism in the mass citizen army. With the rise of nationalism after the French Revolution, European states switched from maintaining a mercenary army (incorporating a large proportion of foreigners) to an army drawn solely from its citizens. From this time forward, every able-bodied male citizen was a potential soldier and the whole nation was turned into a war machine. The scale of warfare could hardly increase any further. For instance, in the autumn of 1794, the size of the French army was reported to be 1,169,000, though its real size was probably 730,000, the largest army France had fielded to that time.[231] McNeill (1982) observes:

> "This figure (the size of the French army at July 1793) was more than twice that Louis XIV had ever been able to put into uniform. Doubling of the army's size (on the basis of a population only about 30 percent greater in 1789 than in 1700) offers rough measure of the intensification of mobilization for war that the revolution wrought in France." (197).

Other European great powers at war with France evinced a similar increase in army size soon after amounting to total war.[232]

The expectation, preparation and experience of wars led states to enlarge their role in the economy, and that role came to include welfare provision.[233] Total warfare not only required a healthy population and contented workers for the war effort, it also needed a population that could reproduce abundantly to replenish the immense human losses the war generated. That brought in the modern welfare state, an example of which is Bismarkian social imperialism. Soldiers, when on reserve, were workers, farmers or other professionals and thereby contributed to the economy in peacetime and to the military during war.[234] The portion of the population not directly involved in combat was important to the military effort too, as they sustained the wartime economies and contributed to the logistic effort. As the masses were mobilized and organized for war and the industrial economy, their welfare became a major concern. Consequently, European states expanded their welfare programs to care for those now deeply involved in the war effort both on and off the field. A population well-trained and taken care of better supported the military effort and the industrial economy. Furthermore, such a population was less likely to cause civil or industrial unrest or be attracted to radical ideologies

---

231 Refer to Best (1982) and Black (1991, pp. 625).
232 Refer to McNeill (1982, Ch. 6) and Cohen (1985, Chh. 2, 3).
233 Refer to Bird (1971) for the expanding role of the state in the economy.
234 Refer Van Creveld (1977) and Cohen (1985).

and political actions.[235] They became more willing to make sacrifices for the purpose of the war effort.[236]

Comparison of 19th century European military forces with those of the rest of the world reveals a great discrepancy. The increasingly larger mass armies of Europe, primarily made up of conscripts and reserves, came to dominate the military scene. However, such large citizen armies were not for export to the other parts of the world. Asian rulers could not create such a structure, being too weak to either create or maintain armed forces on such a scale. More importantly, in Africa and Asia, states could not trust the general populace to not attack its rulers if it had the chance.[237]

From 1840 onwards, the industrial revolution influenced how war was conducted as war became industrialized.[238] The use of steam engines, steel hull ships and the railway increased transportation capacity. Machines replaced artisans in the mass production of arms. Arms production was now achieved speedily and in large volumes to supply the ever-growing mass armed forces.[239] A global, industrialized armaments business emerged in the 1860s.[240] The increased scale of warfare and industrialization of war in turn caused the rise of command technology: innovation was directed and planned with an eye on its military implications. States began to pour resources into education, research and development and changes in weapon design accelerated, with greater technological spillovers into the civilian sector.[241] This led to the industrialization of the military, as well as the militarization of industry and society. The state also assumed a greater role in industry.[242] An extreme manifestation of the increasing militarization of society and state was the rise of the highly militarized totalitarian states of Stalinist Soviet Union, Nazi Germany and Fascist Italy.

Industrial, scientific and technological communities related to the military became the pioneers of new technology. These new technologies included advances in steel metallurgy, industrial chemistry, electrical machinery, radio communication, turbines, fuels, optics, calculators and hydraulic machinery. The military-industrial complex quickly evolved into vast bureaucratic structures of a quasipublic character; and decisions of the big arms firms, whether technical or financial, began to assume public importance. The actual quality

---

235 Refer to Goldstone (1991).
236 Refer to Porter (1994, pp. 158161, 179192, 236239).
237 Refer to McNeill (1982, pp. 2567).
238 Refer to Nef (1950), McNeill (1982), Lynn (1993), Weiss and Hobson (1995, Ch. 4) and Bousquet (2009).
239 Refer to McNeill (1982, pp. 2356).
240 Refer to McNeill (1982, p. 241) and Kennedy (1987).
241 Refer to McNeill (1982, pp. 2378, 2924).
242 Refer to Gerschenkron (1962) and Porter (1994).

of weapons produced by these firms mattered vitally to the rival states and armed services of Europe. Following the wars of German unification, states recognized that technical superiority might bring decisive advantage in wars. Each technical option in arms design carried a heavy freight of political and military implications. Scientific and industrial decisions of the firm concerned both the national interest and the financial future of the arms firm.[243] War had become truly industrialized as industry became no less truly militarized.[244]

Internal security and stability were very important for both war mobilization efforts and industrialization; organization and mobilization of the homeland and occupied territories for the war effort and industrial production required a great policing capacity. The rise of radical ideologies and unconventional warfare accentuated the need for policing capacity. The police force undertook new functions at home and abroad to facilitate the war effort and industrial production, including intelligence collection, counter-intelligence, subversion, surveillance, and the monitoring and manipulation of the population's political thinking and inclinations. Effectively, a professional, salaried and bureaucratic police force, specialized in the control of the civilian population, was established, thereby freeing armies to concentrate on external conquests and international wars. Initially, the army had done most of the work of internal control, but police forces made armies better able to focus on external threats.[245] In sum, to achieve greater national power, the state industrialized the military, militarized the economy and politicized the society.[246]

The transformation the European states went through placed them far ahead of the other major Eurasian cultures in all fields of human achievements, most notably in the possession of overseas colonies and trading opportunities.[247] Given their immense superiority, militarily and otherwise, every corner of the globe was turned into a stage for the Europeans to play out their power games during the 19th century. McNeill (1999) observes that in Great Britain in the latter part of the nineteenth century there was:

"...... a new "hard boiled" school of thought which considered the survival of the fittest to be the key to all history and human life, and believed that the future of the Anglo-Saxon race depended on how much of the earth they could seize and settle with British colonists." (480)

---

243 Refer to McNeill (1982, p. 292) and Weiss and Hobson (1995).
244 Refer to McNeill (1982, pp. 3589) and Keegan (1993).
245 Refer to Chapman (1970), Liang (1992) and Tilly (1992).
246 Refer to Clarkson and Cochran (1941), Nef (1950), Ardant (1975), McNeill (1982, Chh. 710), Rasler andThompson (1989), Tilly (1992), Porter (1994) and Weiss and Hobson (1995).
247 Refer to Manning and McMillan (1979), Ishizawa (1988), Clarida and Findlay (1991), Bhagwati, Panagariya and Wong (1995), Thompson and Modelski (1996), Durkin (1997) and Srinivasan (1998).

Almost unnoticed by the European populace, a series of colonial wars resulted in the significant expansion of the European colonial empires, especially those of Britain, France, and Russia (which expanded eastwards across the Eurasian continent to the Pacific Ocean) in both Africa and Asia.

The industrial revolution and the associated transformation of European states, economies and societies enabled European powers to conduct war with greater efficiency and on a larger scale. Enemies hereto impossible to defeat or subdue could now be easily and swiftly dealt with. The most prominent of these enemies were the two great far flung East Asian powers, unchallenged by Western powers, even in the early nineteenth century, namely China and Japan.[248] The economic might and logistical capacity conferred by the industrial revolution upon the Western nations made it possible for them to challenge both China and Japan. The First Opium War between Britain and China ended the self-imposed isolation of China, when a small British detachment swiftly crushed the forces available to the Chinese emperor. This was the most important demonstration of the superior strength of the European nations. There was from then on, only one geopolitical or international system and it was centered in Europe.[249]

This greater European war capacity rested not merely with more advanced military technology, but with the greater fiscal capacity of the modern state and ultimately, the greater economic might of an industrial society. Qing Dynasty China modernized her military, but not her fiscal regime or economy, and was defeated repeatedly by Japan and the Western powers. On the other hand, after the gunboat diplomacy of Perry ended the isolation of Japan, Japan westernized completely and modernized its political, fiscal, economic and social systems, besides of course its military system. Japan achieved parity in war efficiency with the Western powers soon after Japanese reforms started in 1870s, and was able to defeat China in 1894 and Russia in 1905.

6. Conclusions

A cross-sectional examination of the major Eurasian cultures reveals that the gunpowder empires were all stable and secure regimes that controlled an extensive territory or sphere of influence. Within their geopolitical niche, they were practically uncontestable and had nothing to fear, and as a result became conservative, complacent and lethargic revenue pumps. Consequently, the

---

248 Refer to Parker (1996, pp. 153–154) and McNeill (1999, Part III).
249 Refer to Porter (1994, pp. 1467).

developmental momentum of these diverse lands, a momentum that had accumulated under the medieval fragmented and decentralized political environment, died out. These gunpowder empires remained largely stagnant until forcefully incorporated into the Europe-centered modern world state system. Long periods of peace in their international environment caused the failure of China, Japan, and the other gunpowder empires to produce modern states.[250] Asian trade was shrinking throughout these centuries and the gunpowder empires in synchronized decline.[251] Elvin (2004) describes the puzzle this way:

> "Why did major technological creativity fade away in late-imperial China (though skilled fine-tuning and small inventions did not vanish)? Why did the great creativity in medieval Islam, exemplified by Ibn al-Haytham (Alhazen) in optical science and Ibn Khaldun, apparently die away in later centuries?" (34-35)

In sum, the gunpowder military technological revolution significantly affected geopolitics and economic performance of the major Eurasian cultures. The gunpowder military revolution caused a significant increase in the economies of scale in warfare and led to the emergence of gigantic continental size empires in the major oriental Eurasian cultures each with a dominant core area. In contrast, in Europe with her fragmented and multiple core areas geography, the state system was preserved and the gunpowder revolution served mainly to intensify political military competition between the European states. Differences in political military competition brought about by the gunpowder military revolution explains the splendid and explosive advances of the pluralistic and competitive European system and the dull and dismal performance of the oriental major gunpowder empires.

---

250 Refer to Parker (1996).
251 Refer to Wesson (1967, 1978), Simkin (1968, pp. 258-9) and Jones (1981, p. 170; 1988).

EPILOGUE

The Contemporary Global State System

The competitive state system of Europe propelled European nations to achieve two major feats in world history: one was the industrial revolution, and the other was the creation of a western-centered global state system. The industrial revolution significantly shifted the relative combined military and economic efficiency to the advantage of the western nations. Military capacity between the western industrialized nations and the rest of the world became extremely asymmetric. Furthermore, the industrial revolution greatly increased the logistical capacity of the industrialized nations and their military. Western great powers possessed far greater capability to project power overseas due to the industrial revolution. The industrialization and mechanization of warfare caused the mass factor to become larger. The outcome was an extremely high concentration of resources and capability in the hands of the western industrial nations.

Conquests of far-flung corners of the world became possible given the greater logistical capacity and the larger mass factor. The whole globe was subjected to political and military penetration by the western industrialized nations.[252] Even the gunpowder empires of the major oriental Eurasian cultures found it hard to resist encroachments from the West after the industrial revolution. Of course the other major cultures responded and resisted, yet their responses and resistance were mostly unsuccessful, especially initially, given the sudden drastic changes in relative combined military and economic efficiency due to the industrial revolution. The only notable exception was Japan.

The fragmented geography of Japan and high contestability of the Tokugawa Shogunate meant that Japan, unlike the other major non-European Eurasian cultures, had sunk least into the slumber of gunpowder imperial complacency and conservatism. Cultures, customs, institutions and policies were less at odds with the requirements of a modern state, economy and society. This was especially so given that there were still many feudal lords in Japan that were highly independent of the Tokugawa Shogunate. These feudal lords were mainly concentrated in southwestern Japan. A spirit of competition still survived in the Japanese system. Therefore, once there was a political will to modernize, there were less frictions and obstacles to obstruct the effort and hence the

---

252 Refer to Parker (1995).

move for reforms was easier for Japan than for the other major oriental cultures. Furthermore, the relatively more fragmented Japanese political system during the Tokugawa Shogunate also meant that it was much easier for the reformed minded regional lords to challenge the power of the complacent and conservative Tokugawa Shogunate and take over the government and steer Japan on a course of reform. Consequently, Japan responded to the challenge of the Western great powers decisively and rationally by undertaking a drastic westernization program, while the responses of the other major gunpowder empires were much more lukewarm.

Another important reason for the early failure of the other major cultures in responding to challenges from the West during this period was the size of the mass factor. A large mass factor confers great advantage to the more efficient and stronger contestant, and the weaker rival thereby has a smaller chance of victory. A larger capability is thus transformed into a great advantage in conflicts, and any defeat could be decisive. Consequently, India was colonized and the Ottoman Empire, Persia and China lost much of their territory and sphere of influence. Even Japan, the only major culture that succeeded in meeting the Western challenge by full-heartedly and completely westernizing itself, suffered defeats and humiliations during her initial contacts with the Western powers.

The failure of the rest in resisting Western encroachments and colonization turned international politics into an internal struggle between Western nations. Yet, the intense political-military competition among the Western great powers culminated in two world wars which weakened the Western nations and helped to end the Western dominance over the rest. The global state system that the West created was transformed into a westernized global state system.

There were important military technological changes that led to the retreat of Western powers and the political independence of the rest, as well as a resurgence of the older major cultures. Two important macro historical forces were at work here. The first was the decreasing size of the mass factor. The consequence was a more symmetrical distribution of resources and capability. In the language of international relations scholars, it was a move towards the multi-polar world order. In the language of macro historians, it was the coming of a new medieval era. The above mega-trend allowed the next mega-trend time to work itself out. The next mega-trend was the diffusion of knowledge across civilizations and nations.[253] This process led to a more symmetrical distribution of relative combined military and economic efficiency. This mega-trend further reinforced the movement toward a more symmetrical distribution of resources and capability and a multi-polar world order initiated by a small mass factor.

---

253 Refer to McNeill (1963, 1998).

The rise of mass politics and its associated guerrilla warfare and the development of nuclear weapons caused the reduction in the size of the mass factor. These military technological changes also lead to a more symmetrical relative combined military and economic efficiency between the advanced nations and the developing world. The industrial revolution brought forth many new products, including the invention and use of mass media and mass printing. These inventions released the power of the general public. Mass involvement in politics and warfare made war total. The battleground between advanced powers and backward societies was to some extent leveled. With mass involvement and guerrilla warfare, international conflicts became prolonged and less decisive. Economies of scale in conflicts became smaller.

Given a smaller mass factor, the western great powers, despite their advanced technology, better organization and larger resource base, lost part of their strategic advantage. With the loss of parts of their military and political advantage, the western powers retreated, especially after World War II. The rise of communist China under the leadership of a totalitarian mass party and the independence of India and other Asian states through mass popular resistance movements testified to the coming of a new political-military macro-historical era. The invention of nuclear weapons further leveled the playing field between advanced and less developed nations. With nuclear weapons, it became almost impossible to attack and conquer another nuclear power, no matter how backward it is. Consequently, communist China, despite its international isolation, policy failures, and poverty, retained its independence and integrity throughout the Cold War. Mass politics won India its independence, something that India had not enjoyed for centuries—not since the Mughal Empire fell to the British colonial regime. Other smaller nations benefited from the smaller economies of scale in conflict as well. For instance, Vietnam defeated France and America to regain independence and unification. The military-political preconditions for the resurgence of the non-western world were laid down.

The above restructuring in international politics created a more even distribution of capability. The more even distribution of capability energized the independent non-western nations to greater developmental effort. Through trial and error and experimenting with different development strategies, the non-western states have slowly acquired a share of the world economy more proportionate to their national potential. After centuries of decline and stagnation, the Asian economies have boomed again in recent decades. First it was Japan who industrialized in late nineteenth century after Meiji Restoration and rebuilt herself after the destruction of World War II. Then the newly industrialized economies of South Korea, Taiwan, Hong Kong and Singapore emerged after World War II. Then it was the second wave of newly industrialized economies of Malaysia, Thailand and Vietnam, and the giant economies of China and India. By projection, within decades, Asia will regain its historical preeminence in the world economy.[254] Other non-western nations forge ahead as well, most notably Brazil. Rising along with the rapid growth of the Chinese, Indian, Brazilian and other non-western economies is their overall national power, including military power.[255]

The resurgences of China and India are especially illuminating on how international political-military competition spurs economic development. The need to cope with the threat from the former Soviet Union and the United States, and to regain its traditional preeminence in the international arena, prompted Deng Xiaoping's reforms in China. Isolationist and socialist policies were abandoned in favor of market-oriented reforms and open door policies. The Chinese economy took off, and China's economic rise has increased its overall national power. This alarmed the surrounding nations, including the other Asian giant, India. India responded to the economic rise of China by starting its own economic reforms, ending her self-sufficiency policy and socialist measures in favor of greater integration with the world economy and market-friendly policies. All these were adopted with the goal of faster economic growth and greater national power. Other economies in the region have followed suit as well. In sum, these major non-western nations are competitively mobilizing economic resources for greater capability in the international arena, as happened repetitively in world history.

An important technological change that has and is still shaping the contemporary global state system is the computer technology. It decreases the economies of scale in information technology and production which leads to the downsizing of firms and the decline of American manufacturing. It also leads to the rise of smaller firms from Japan and the small, East Asian newly industrializing economies of Korea, Taiwan, Hong Kong and Singapore as well as other late comers in industrialization. Another result it brings is a decline of the mass factor. It helps in the miniaturization of weapons, especially weapons with mass destruction capability. The decreasing mass factor due to changes in current military technology together with the decreasing scale of production and information technology increases the difficult of state building and maintenance and state resource extraction and led to increasing number of non state force wielding organizations. This might explains the increasing cases of state failure in the contemporary world. Therefore, the big question confronting us is - will there be a new dark age in the future, as one that followed the collapse of the Roman Empire?[256] And what would be the economic consequences of these technological changes? The analytical concepts exposed in this book will definitely be useful in the search for an answer.

---

254 Refer to Pomeranz (2000).
255 Refer to Samuels (1994), Cohen (2001), Feignebaum (2003) and Ganguly (2003).
256 Refer to Bonomo, Bergamo, Frelinger, Gordon IV and Jackson (2007).

Bibliography

AbuLughod, J. L. (1989). *Before European Hegemony: The World System A.D. 1250-1350*. New York: Oxford University Press.

Acemoglu, D. and Robinson, J. A. (2012). *Why Nations Fail: The Origins of Power, Prosperity and Poverty*, New York: Crown Business.

Alesina, A., Spolaore, E. and Wacziarg, R. (2000). Economic Integration and Political Organization. *The American Economic Review* 90 (5), 1276-1296.

Anderson, G. M. (1992). Cannons, Castles, and Capitalism: the Invention of Gunpowder and the Rise of the West. *Defense Economics*, (3), 147-160.

Anderson, J. E. and Marcouiller, D. (2005). Anarchy and Autarky: Endogenous Predation as a Barrier to Trade. *International Economic Review* 46, (1), 189-213.

Anderson, P. (1975). *Lineages of the Absolutist State*. London: New Left Books.

Ardant, G. (1975). Financial policy and economic infrastructure of modern states and nations. In C. Tilly ed., *The Formation of the National States in Western Europe*, (pp. 164-242). Princeton: Princeton University Press.

Armstrong, K. (2000). *Islam: A Short History*. New York: Modern Library.

Bacon, F. (1972). *Essays. (Introduction by Michael J. Hawkins)*. London: J. M. Dent & Sons.

Baechler, J. (1976). *The Origins of Capitalism*. New York: St. Martin's Press.

Baechler, J., Hall, J. A. and Mann, M. (Eds.). (1988). *Europe and the Rise of Capitalism*. Oxford: Basil Blackwell Ltd.

Baldwin, D. A. (Ed.). (1993). *Neorealism and Neoliberalism: the Contemporary Debate*. New York: Columbia University Press.

Baran, P. A. (1957). *The Political Economy of Growth*. New York: Monthly Review Press.

Baron, H. (1955). *The Crisis of the Early Italian Renaissance*. Princeton, NJ: Princeton University Press.

Barro, R. J. and Sala-i-Martin, X. (1998). *Economic Growth*. Cambridge, MA: MIT Press.

Barzel, Y. (1989). *Economic Analysis of Property Rights*. Cambridge: Cambridge University Press.

Bean, R. (1973). *War and the birth of the nation state*. Journal of Economic History 33,203-221.

Bentley, J. H. and Adas, M. (Eds.). (1995). *Shapes of World History in Twentieth-Century Scholarship*. Washington, DC: American Historical Association.

Berinstain, V. (1998). *Mughal India: Splendours of the Peacock Throne*. London: Thames & Hudson Ltd.

Berman, H. J. (1983). *Law and Revolution: The Formation of the Western Legal Tradition*. Cambridge, MA: Harvard University Press.

Berman, H. J. (2003). *Law and Revolution II: The Impact of the Protestant Reformations on the Western Legal Tradition*. Cambridge, MA: The Belknap Press of Harvard University Press.

Bernholz, (1998). International Competition Among States: Institutions, Market Regime and Innovations in Antiquity: A comment and An Extension to the Cases of Sumer and Phoenicia. In P. Bernholz, M. E. Streit and R. Vaubel (Eds.), *Political Competition, Innovation and Growth: A History Analysis*. (pp. 109-125). Berlin: Springer-Verlag.

Bernholz, P., Streit, M. E. and Vaubel, R. (Eds.).(1998). *Political Competition, Innovation and Growth: A Historical Analysis*. Berlin: Springer-Verlag.

Bernholz, P. and Vaubel, R. (Eds.). (2004). *Political Competition, Innovation and Growth in the History of Asian Civilizations*. Cheltenham, UK: Edward Elgar.

Bernstein, R. and Munro, R. (1994). *The Coming Conflict with China*, Random House, New York.

Best, G. (1982). *War and Society in Revolutionary Europe, 1770-1870.* New York: St. Martin's Press.

Bhagwati, J. N., Panagariya, A. and Srinivasan, T. N. (1998). *Lectures on International Trade.* Cambridge, MA: The MIT Press.

Bird, R. M. (1971). *Wagner's law of expanding state activity.* Public Finance 26, 126.

Black, J. (1991). *A Military Revolution? Military Change and European Society 1550-1800.* Atlantic Highlands, NJ: Humanities Press.

Blum, U. (1991). A spatial model of the state. *Journal of Institutional and Theoretical Economics* 147, (2), 312-36.

Blum, U. and Dudley, L. (1989). A spatial approach to structure change the making of the French hexagon. *Journal of Economic History* 49, 657-675.

Blum, U. and Dudley, L. (2003). Standardised Latin and medieval economic growth, *European Review of Economic History* 7, 213-238.

Bodde, D. and Fung, Y. L. (1983). *A History of Chinese Philosophy.* Princeton, NJ: Princeton University Press.

Bonomo, J., Bergamo, G., Frelinger, D. R., Gordon IV, J. and Jackson, B. A. (2007). *Stealing the Sword: Limiting Terrorist Use of Advanced Conventional Weapons,* Santa Monica, CA: Rand Corporation.

Bousquet, A. (2009). *The Scientific Way of Warfare: Order and Chaos on the Battlefields of Modernity,* New York: Columbia University Press.

Braudel, F. (1949). *The Mediterranean and the Mediterranean World in the Age of Philip II.* London: Collins.

Breton, A. (1989). The growth of competitive government. *Canadian Journal of Economics* 22, 717-750.

Brewer, J. (1989). *The Sinews of Power: War, Money, and the English State, 1688-1783.* NewYork: A. A. Knopf.

Bronfenbrenner, M. (1964). The appeal of confiscation in economic development. In O. Feinstein (Ed.), *Two Worlds of Change*, Garden City, New York: Anchor Books.

Burke, E. (Ed.). (1993). *Rethinking World History: Essays on Europe, Islam, and World History. Cambridge*: Cambridge University Press.

Bush, W. C. (1974). Some implications of anarchy for the distribution of property. *Journal of Economic Theory* 8, 401-412.

Buzan, B. and Little, R. (2000). *International Systems in World History: Remaking the Study of International Relations*. Oxford: Oxford University Press.

Cahen (1970). In Holt, P. M, Lambton, A. K. S and Lewis, B. (Eds.), The Cambridge *History of Islam 2*. Cambridge: Cambridge University Press. (511-38).

Chapman, B. (1970). *Police State*. London: Pall Mall.

Chaudhuri, K. N. (1990) *Asia Before Europe: economy and civilisation of the Indian Ocean from the rise of Islam to 1750*. Cambridge: Cambridge University Press.

Chernow, Ron. (2004). *Alexander Hamilton*. New York: Penguin Press.

Church, P. (2006). *A Short History of Southeast Asia*. 4th Ed. Singapore: John Wiley and Sons.

Cipolla, C. M. (1966). *Guns, Sails and Empires: Technological Innovation and European Expansion 1400-1700*. Barnes and Noble Books.

Cipolla, C. M. (1967). *Clocks and Culture 1300-1700*. London: Collins.

Clarida, R. H. and Findlay, R. (1991). Endogenous Comparative Advantage, Government and the Pattern of Trade. *NBER working paper* no. 3813.

Clarkson, J. D. and Cochran, T. C. (1941). *War as a Social Institution*. New York: Columbia University Press.

Coase, R. H. (1937). The nature of the firm. *Economica 386*.

Coase, R. H. (1960) The problem of social cost. *Journal of Law and Economics*.

Coase, R. H. (1988). *The Firm, the Market, and the Law*. Chicago: University of Chicago Press.

Cohen, E. A. (1985). *Citizens and Soldiers: The Dilemmas of Military Service.* Ithaca: Cornell University Press.

Cohen, S. P. (2001). *India: Emerging Power,* Washington D. C.: Brookings Institution.

Cohen, Y., Brown, B. R. and Organski, A. F. K. (1981). The paradoxical nature of state making: the violent creation of order. *American Political Science Review* 75: 901-910.

Coleman, D. (1961). Economic problems and policies. In *New Cambridge Modern History.* Vol. 5, Cambridge: Cambridge University Press.

Cook, Michael, (2004). Comments in Bernholz, P. and Vaubel, R. (Eds.) (2004). *Political Competition, Innovation and Growth in the History of Asian Civilizations.* (pp. 29-30). Cheltenham, UK: Edward Elgar.

Cowen, T. (1990). *Economic effects of a conflictprone world order.* Public Choice 64, 121-134.

Crosby, A. W. (1972). *The Columbian Exchange: Biological and Cultural Consequences of 1492.* Westport, CN:

Crosby, A. W. (1986). *Ecological Imperialism: The Biological Expansion of Europe, 900-1900.* Cambridge: Cambridge University Press.

Crosby, A. W. (1994). *Germs, Seeds and Animals: Studies in Ecological History.* New York: M. E. Sharpe.

Davis, R. (1965). The rise of protection in England, 1689-1786. *Economic History Review 2,* (19), 306-317.

Davis, R. W. (Ed.). (1995). *The Origins of Modern Freedom in the West.* Stanford, CA: Stanford University Press.

Deng, K. G. (2000). A critical survey of recent research in Chinese economic history. *Economic History Review 53,* (1), 1-28.

De Vries, J. (1998). Innovation and growth in the Netherlands: searching for explanations. In P. Bernhozl, M. E. Streit and R. Vaubel (Eds.), *Political Competition, Innovation and Growth: A History Analysis.* (209-221). Berlin: Springer-Verlag.

Diamond, J. M. (1997). *Guns, Germs, and Steel: the Fates of Human Societies*: New York: W. W. Norton.

Distelrath, G. (2004). Advantages of centralized and decentralized rule in Japan. In Bernholz, P. and Vaubel, R. (Eds.). *Political Competition, Innovation and Growth in the History of Asian Civilizations*. Cheltenham, UK: Edward Elgar. (96-112).

Downing, B. M. (1992). *The Military Revolution and Political Change: Origins of Democracy and Autocracy in Early Modern Europe*. Princeton, NJ: Princeton University Press.

D'souza, Rohan. (2002). Crisis before the fall: some speculations on the decline of the Ottomans, Safavids and Mughals. *Social Scientist* 30, (9/10), 3-30.

Dudley, L. (1990). Structural change in interdependent bureaucracies: was Rome's failure economic or military? *Explorations in Economic History* 27, 232-248.

Dudley, L. (1991). *The Word and the Sword: How Techniques of Information and Violence Have Shaped Our World*. Cambridge, MA: Basil Blackwell.

Dudley, L. (1992). *Punishment, reward and the fortunes of states*. Public Choice 74, 293.

Duffy, M. (Ed.). (1980). *The Military Revolution and the State, 1500-1800*. Exeter: University of Exeter. Exeter Studies in History.

Durkin, J. T. (1997). Perfect competition and endogenous comparative advantage. *Review of International Economics* 5, (3), 401-411.

Eberhard, W. (1960). *A History of China*. London: Routledge and Kegan Paul.

Elvin, M. (1973). *The Pattern of the Chinese Past*. London: Eyre Methuen.

Elvin, M. (2004). Comments in Bernholz, P. and Vaubel, R. (Eds.) (2004). *Political Competition, Innovation and Growth in the History of Asian Civilizations* (pp. 34-35). Cheltenham, UK: Edward Elgar.

Evans, P., Rueschemeyer, D. and Skocpol, T. (Eds.). (1985). *Bringing the State Back in*. Cambridge: University of Cambridge Press.

Findlay, R. and O'Rourke, K. H. (2007). *Power and Plenty: Trade, War, and the World Economy in the Second Millennium*, Princeton, New Jersey, Princeton University Press.

Filesi, T. (1972). *China and Africa in the Middle Ages.* London: Frank Cass.

Fitzgerald, C. P. (1973). *China and Southeast Asia Since 1945.* London: Longman.

Foran, J. (1992). The long fall of the Safavid Dynasty: Moving beyond the standard views. *International Journal of Middle East Studies 24,* (2), 281-304.

Frank, A. G. (1975). *On Capitalist Underdevelopment.* New York: Oxford University Press.

Frank, A. G. (1998). *ReOrient: Global Economy in the Asian Age.* University of California Press.

Frank, A. G. and Gills, B. K. (Eds.). (1993). *The World System: Five Hundred Years or Five Thousand?* London: Routledge.

Friedman, D. (1977). A theory of size and shape of nations. *Journal of Political Economy 85,* (1), 5977.

Friedman, D. (1979). Private creation and enforcement of law: a historical case. *Journal of Legal Studies 8,* 399-415.

Fung, Y. L. (1966). *A Short History of Chinese Philosophy.* Free Press.

Ganguly, S. (ed.) *India as an Emerging Power,* London: Frank Cass Publishers.

Gardiner, P. ed. (1959). *Theories of History.* New York: Free Press.

Garfinkel, M. R. (2004). Stable Alliance Formation in Distributional Conflict. *European Journal of Political Economy 20,* (4), 829-852.

Garthwaite, G. R. (2005). *The Persians,* Blackwell Publishing, Malden, MA, USA.

Gerschenkron, A. (1962). *Economic Backwardness in Historical Perspective.* Cambridge, MA: Belknap Press.

Gibbon, E. (1932). *The Decline and Fall of the Roman Empire.* New York: Modern Library.

Gilpin, R. (1981). *War and Change in World Politics.* Cambridge: Cambridge University Press.

Gimpel, J. (1977). *The Medieval Machine.* London: Gollanz.

Goemans, H. (2000). *War and Punishment: The Causes of War Termination and the First World War.* Princeton, N. J.: Princeton University Press.

Goldstone, J. A. (1991). *Revolution and Rebellion in the Early Modern World.* University of California Press.

Goldschmidt, A. Jr. (2002). *A Concise History of the Middle East.* 7th Ed. Westview Press.

Gowa, J. (1989). Bipolarity, multipolarity, and free trade. *American Political Science Review 83,* 1245-1256.

Gowa, J. (1994). *Allies, Adversaries, and International trade.* Princeton, NJ: Princeton University Press.

Gowa, J. and Mansfield, E. D. (1993). Power politics and international trade. *American Political Science Review 87,* 408-420.

Graham, A. C. (1973). China, Europe and the origins of modern science: Needham's *The Grand Tradition,* in Chinese Science: *Explorations of an Ancient Tradition,* S. Nakayama and N. Sivin (Eds.)., 45-69, Cambridge, MA: MIT Press.

Grieco, J. M. (1988a). Realist theory and the problem of international cooperation: analysis with an amended prisoner's dilemma model. *Journal of Politics 50,* (3), 600-624.

Grieco, J. M. (1988b). Anarchy and the limits of cooperation: a realist critique of the newest liberal institutionalism. *International Organization 42,* (3), 485-507.

Grieco, J. M. (1990). *Cooperation among Nations: Europe, America and Non-tariff Barriers to Trade.* Ithaca: Cornell University Press.

Grieco, J. M., Powell, R., and Snidal, D. (1993). The relative-gains problem for international cooperation. *American Political Science Review 87,* (3), 729-743.

Grousset, R. (1970). *The Empires of the Steppes: A History of Central Asia.* Rutgers University Press.

Haakonssen, K. ed. (1994), *Political Essays.* Cambridge: Cambridge University Press.

Hale, O. (1971). *The Great Illusion, 1910-1914.* New York: Harper and Row.

Harrison, J. A. (1972). *The Chinese Empire: A Short History of China from Neolithic times to the end of the Eighteenth Century*. New York: Harcourt Brace Jovanovich.

Hartwell, R. M. (1966). Markets, technology, and the structure of enterprise in the development of the eleventh century Chinese iron and steel industry. *Journal of Economic History 26*, 29-58.

Hartwell, R. M. (1998). Comment. In Bernholz et al., *Political Competition, Innovation and Growth: A History Analysis*. (pp. 239-241). Berlin: Springer-Verlag.

Haywood, J. (1997). *The Complete Atlas of World History: Prehistory and the Ancient World,4,000,000 Years Ago - AD 600*. Oxford: Sharpe Reference.

Head, J. W. and Wang, Y. (2005). *Law Codes in Dynastic China: A Synopsis of Chinese Legal History in the Thirty Centuries from Zhou to Qing*. Durham, NC: Carolina Academic Press.

Hechter, M. (1980). Regional modes of production and patterns of state formation in WesternEurope. *American Journal of Sociology 85*.

Heckscher, E. (1955). *Mercantilism. Vol. 2*. London: Allen and Unwin.

Heidhues, M. S. (2000). *Southeast Asia: A Concise History*. London: Thames and Hudson.

Hicks, J. (1969). *A Theory of Economic History*. Oxford: Clarendon Press.

Hintze, O. (1975). Military Organization and the Organization of the State. In F. Gilbert (Ed.), *The Historical Essays of Otto Hintze*. New York: Oxford University Press.

Hirshleifer, J. (1988). *The analytics of continuing conflict. Synthese 76*, 201-233.

Hirshleifer, J. (1989). *Conflict and rentseeking success functions: ratio vs. difference models of relative success*. Public Choice 63: 101-112.

Hirshleifer, J. (1991). The technology of conflicts as an economic activity. *Journal of Economic Perspectives 81*, (2), 130-134.

Hirshleifer, J. (1995). Anarchy and its Breakdown. *Journal of Political Economy 103*, 1, 26-52.

Hirshleifer, J. (2000). The Macrotechnology of Conflict. *Journal of Conflict Resolution 44*, (6),773-792.

Hirshleifer, J. (2001). *The Dark Side of the Force: Economic Foundations of Conflict Theory*. Cambridge: Cambridge University Press.

Holt, P. M, Lambton, A. K. S and Lewis, B. (Eds.). (1970). *The Cambridge History of Islam 2*. Cambridge: Cambridge University Press.

Huang, R. (1974). *Taxation and Governmental Finance in Sixteenth-Century Ming China*. London: Cambridge University Press.

Huang, R. (1981). *1587 : A Year of No Significance - the Ming Dynasty in Decline*. New Haven, CN: Yale University Press.

Huang, R. (1988). *China, A Macro History*. New York: East Gate.

Huff, T. E. (2004). Comments. In Bernholz, P. and Vaubel, R. (Eds.). *Political Competition, Innovation and Growth in the History of Asian Civilizations*. (204-210). Cheltenham, UK: Edward Elgar.

Hui, V. (2005). *War and State Formation in Ancient China and Early Modern Europe*. Cambridge: Cambridge University Press.

Ishizawa, S. (1988). Increasing returns, public inputs, and international trade. *The American Economic Review 78*, 794-795.

Jansen, M. B. (2002). *The Making of Modern Japan*. Harvard University Press.

Jaspers, Karl. (1953). *The Origin and Goal of History*. (M. Bullock, Trans.). New Haven, CN: (Original work published in 1949).

Jones, E. L. (1974). Institutional determinism and the rise of the Western world. *Economic Inquiry 12*,114-124.

Jones, E. L. (1981). *The European Miracle: Environments, Economies and Geopolitics in the History of Europe and Asia*. Cambridge: Cambridge University Press.

Jones, E. L. (1988). *Growth Recurring: Economic Change in World History*. Oxford: Clarendon Press.

Jones, E. L. (1990). The Real Question about Chinese History: Why was the Song Economic Achievement Not Repeated? *Australian Economic History Review 30*, (2), 5-22.

Jones, E. L. (2002). *The Record of Global Economic Development*. Northampton, MA: Edward Elgar Publishing.

Kavolis, V. (1972). *History on Art's Side: Social Dynamics in Artistic Efflorescence*. Ithaca: Cornell University Press.

Keegan, J. (1993). *A History of Warfare*. Vintage Books, New York.

Kempin, Frederick G. (1990). *Historical Introduction to Anglo-American Law in a Nutshell*. St. Paul, MN: West Group.

Kennedy, P. (1987). *The Rise and Fall of the Great Powers*. Random House.

Khan, I. A. (2004). *Gunpowder and Firearms: Warfare in Medieval India*. Oxford: Oxford University Press.

Kulke, H. and Rothermund, D. (1998). *A History of India*. 3rd Ed. London: Routledge.

Kuran, T. (2004). Islamic statecraft and the Middle East's delayed modernization. In Bernholz, P. and Vaubel, R. (Eds.). *Political Competition, Innovation and Growth in the History of Asian Civilizations*. (153-183). Cheltenham, UK: Edward Elgar.

Lal, D. (2004). India. In Bernholz, P. and Vaubel, R. (Eds.). *Political Competition, Innovation and Growth in the History of Asian Civilizations*. (128-141). Cheltenham, UK: Edward Elgar.

Landes, D. (1969). *The Unbound Prometheus*. Cambridge: Cambridge University Press.

Lee, C. K. (1988). *War in the Confucian International Order*. Unpublished doctoral dissertation, University of Texas, Austin, TX.

Levi, M. (1981). The predatory theory of rule. *Politics and Society 10*, 431-65.

Levi, M. (1988). *Of Rule and Revenue*. Berkeley, CA: University of California Press.

Liang, H. H. (1992). *The Rise of Modern Police and the European State System from Metternich to the Second World War*. Cambridge: Cambridge University Press.

Lynn, J. A., (Ed.). (1993). *Feeding Mars: Logistics in Western Warfare from the Middle Ages to the Present*. Oxford: Westview Press.

Maddison, A. (1971). *Class Structure and Economic Growth: India and Pakistan since the Moghuls*. New York: Norton.

Manning, P. (2003). *Navigating World History: Historians Create A Global Past.* New York.: Houndmills.

Manning, R. and McMillan, J. (1979). Public intermediate goods, production possibilities, and international trade. *Canadian Journal of Economics 12*, (2), 243-257.

Marrese, M. and Va'ous, J. (1983). Unconventional gains from trade. *Journal of Comparative Economics 7*, (4), 382-99.

Martinussen, J. (1997). *Society, State and Market: A Guide to Competing Theories of Development.* London: Zed Press.

McNeill, H. W. (1963). *The Rise of the West: A History of the Human Community.* Chicago: University of Chicago Press.

McNeill, H. W.(1982). *The Pursuit of Power: Technology, Armed Forces, and Society since A.D. 1000.* Chicago: University of Chicago Press.

McNeill, H. W. (1999). *A World History.* 4th Ed. Oxford: Oxford University Press.

Mo, P. H. (2004). Lessons from the history of imperial China. In P. Bernholz and R. Vaubel (Eds.), *Political Competition, Innovation and Growth in the History of Asian Civilizations.* (pp. 57-74). Cheltenham, UK: Edward Elgar

Mokyr, J. (1990). *The Lever of Riches: Technological Creativity and Economic Progress.* Oxford: Oxford University Press.

Moore, B., Jr. (1967). *Social Origins of Dictatorship and Democracy.* London: Allen Lane.

Montesquieu (1748[1989]). *The Spirit of the Laws.* (A. M. Cohler, B. C. Miller and H. S. Stone, Trans.). Cambridge: Cambridge University Press.

Morris, M. F. (1911). *An Introduction to the History of the Development of Law.* Washington, D. C.: John Byrne.

Nadarajah, R. (1992). The link between state organization and military power in late 18th century South India. In M. Dubean and M. Warren, et. al. (Eds.), *South Asian Symposium 1991: A Reader in South Asian Studies*, Toronto: University of Toronto.

Needham, J. (1953). Science and technology in China. In S. Lilley (Ed.), *The Social History of Science.* Copenhagen: Munksgaard.

Needham, J., Ho, P. Y., Lu, D. G. and Wang, L. (1987). *Science and Civilization in China: Volume 5, Chemistry and Chemical Technology, Part 7, Military Technology: the Gunpowder Epic.* Cambridge: Cambridge University Press.

Nef, J. U. (1950). *War and Human Progress: An Essay on the Rise of Industrial Civilization.* Cambridge, MA: Harvard University Press.

North, D. (1981). *Structure and Change in Economic History.* New York: W. W. Norton. North, D. (1987). Institutions, transaction costs and economic growth. *Economic Inquiry 25*, (3), 419-428.

North, D. (1990). *Institutions, Institutional Change and Economic Performance.* Cambridge: Cambridge University Press.

North, D. (1995). The paradox of the West. In R. W. Davis (Ed.), *The Origins of Modern Freedom in the West.* (pp. 7-34). Stanford, CA: Stanford University Press.

North, D. (1998). The rise of the West. In P. Bernhozl, M. E. Streit and R. Vaubel (Eds.), *Political Competition, Innovation and Growth: A History Analysis.* (pp. 13-28). Berlin: Springer-Verlag.

North, D. and Thomas, R. (1973). *The Rise of the Western World: A New Economic History.* Cambridge: Cambridge University Press.

Nti, K. O. (1999). Rent-seeking with asymmetric valuations. *Public Choice 98*, 415-430.

Olson, M. (1993). Dictatorship, Democracy, and Development, *The American Political Science Review 87*, (3), 567-576.

Olson, M. (2000). *Power and Prosperity: Outgrowing Communist and Capitalist Dictatorships*, New York: Basic Books.

Oppenheimer, F. (1975). *The State.* Montreal: Black Rose Books.

Osborne, M. (2000). *Southeast Asia: An Introductory History.* 8th Ed. Allen and Unwin.

Parker, G. (1976). *The military revolution, 1560-1660a myth?* Journal of Modern History.

Parker, G. (1996). *The Military Revolution: Military innovation and the rise of the West, 1500-1800.* Cambridge: Cambridge University Press.

Perkins, D. (1967). Government as an obstacle to industrialization: the case of nineteenth-century China. *Journal of Economic History 27*, 478-92.

Pollard, S. (1998). Political Competition and the British Industrial Revolution. In P. Bernhozl, M. E. Streit and R. Vaubel (Eds.), *Political Competition, Innovation and Growth: A History Analysis.* (pp. 223-237). Berlin: Springer-Verlag.

Pomeranz, K. (2000). *The Great Divergence: Europe, China and the Making of the Modern World Economy.* Princeton, N. J.: Princeton University Press.

Porter, B. D. (1994). *War and the Rise of the State: the Military Foundations of Modern Politics:* New York: Free Press.

Powell, R. (1999). *In the Shadow of Power: States and Strategies in International Politics.* Princeton, NJ: Princeton University Press.

Powelson, J. (2004). Decentralization in the history of Japan. In Bernholz, P. and Vaubel, R. (Eds.). *Political Competition, Innovation and Growth in the History of Asian Civilizations.* (118-127). Cheltenham, UK: Edward Elgar.

Raaflaub, K. and Rosenstein, N., (Eds.). (1999). *War and Society in the Ancient and Medieval Worlds: Asia, The Mediterranean, Europe, and Mesoamerica.* Cambridge, MA: Center for Hellenic Studies, Trustees for Harvard University.

Rasler, K. A. and Thompson, W. R. (1989). *War and State Making: the Shaping of the Global Powers.* London: Unwin Hyman.

Roach, J. (1969). Education and public opinion. In *New Cambridge Modern History.* Vol. 9, Cambridge: Cambridge University Press.

Roberts, M. (1956). *The Military Revolution, 1560-1660.* Belfast.

Rostow, W. W. (1960). *The Stages of Economic Growth.* Cambridge: Cambridge University Press.

Rostow, W. W. (1974). Book review on *The Rise of the Western World: A New Economic History* by D. C. North and R. P. Thomas (1973). *Journal of Economic Literature 12,* (2), 493-496.

Rostow, W. W. (1975). *How it all Began: Origins of the Modern Economy.* New York: McGraw-Hill.

Rothermund, D. (2004). Comment. In P. Bernholz, M. E. Streit and R. Vaubel (Eds.), *Political Competition, Innovation and Growth: A History Analysis.* (pp. 142-146). Berlin: Springer-Verlag.

Ruthven, M. (and Azim Nanji, A). (2004). *Historical Atlas of Islam.* Cambridge, MA: Harvard University Press.

Saggs, H. W. F. (1991). *Civilization before Greece and Rome.* New Haven, CN: Yale University Press.

Samuels, R. J. (1994). *Rich Nation, Strong Army: National Security and the Technological Transformation of Japan.* Ithaca. Cornell University Press.

Sandars, N. K. (1978). *The Sea Peoples: Warriors of the Ancient Mediterranean, 1250-1150 BC.* London: Thames and Hudson.

SarDesai, D. R. (2003). *Southeast Asia: Past and Present.* 5th Ed. Westview Press.

Sawyer, M. C. (1993). *The Seven Military Classics of Ancient China.* Westview Press.

Schumpeter, J. A. (1911). *The Theory of Economic Development.* New York: Oxford University Press.

Simkin, C. G. F. (1968). *The Traditional Trade of Asia.* London: Oxford University Press.

Smith, A. (1776). *An Inquiry into the Nature and Causes of the Wealth of Nations.* University of Chicago Reprint, 1976.

Sonn, T. (2004). *A Brief History of Islam.* Malden, MA: Blackwell Publishing.

Spruyt, H. (1996). *The Sovereign State and Its Competitors: An Analysis of Systems Change,* NJ: Princeton University Press.

Starr, C. G. (1989). *The Influence of Sea Power on Ancient History.* Oxford: Oxford University Press.

Stavrianos, L. S. (1982). *The World to 1500: A Global History.* 3rd edition, Englewood Cliffs, NJ: Prentice-Hall.

Stomberg, A. (1931). *A History of Sweden.* New York: Macmillan.

Stover, L. E. and Stover, T. K. (1976). *China: An Anthropological Perspective.* Pacific Palisades, CA: Goodyear.

Subrahmanyam, S. (1989). Warfare and state finance in Wodeyar Mysore, 172425: a missionary perspective. *Indian Economic and Social History Review* 26, (2).

Tellegen-Couperus, O. (1993). *A Short History of Roman Law.* London: Routledge Taylor & Francis.

Thompson, W. R. and Modelski, G. (1996). *Leading Sectors and World Powers: the Coevolution of Global Politics and Economics.* Columbia, SC: University of South Carolina Press,

Tigar, M. E. and Levy, M. R. (2000). *Law and the Rise of Capitalism.* New York: Monthly Review Press.

Tilly, C. (1975). *The Formation of National States in Western Europe.* Princeton, NJ: Princeton University Press.

Tilly, C. (1992). *Coercion, Capital, and European States, AD 990-1992.* Cambridge, MA: Blackwell.

Toutain, J. (1930). *The Economic Life of the Ancient World.* (M. R. Dobie, Trans.). New York: A. A. Knopf.

Treadgold, W. (2001). *A Concise History of Byzantium.* New York: Palgrave.

Van Creveld, M. L. (1977). *Supplying War: Logistics from Wallenstein to Patton.* Cambridge: Cambridge University Press.

van Klaveren, J. (1969). *General Economic History, 100-1760: From the Roman Empire to the Industrial Revolution.* Munich: Gerhard Kieckens.

Von Ungern-Sternberg, J. (1998). Innovation in Earlyh Greece: The Political Sphere. In P. Bernhozl, M. E. Streit and R. Vaubel (Eds.), *Political Competition, Innovation and Growth: A History Analysis.* (pp. 85-107). Berlin: Springer-Verlag.

Wade, R. (1992). *Governing the Market.* Princeton, NJ: Princeton University Press.

Watson, A. (1974). The Arab agricultural revolution and its diffusion, 700-1100. *Journal of Economic History 34,* 8-35.

Weber, M. (1923[1961]). *General Economic History*. (F. H. Knight, Trans.). New York: Collier Books.

Webber, C. and Wildavsky, A. (1986). *A History of Taxation and Expenditure in the Western World*. New York: Simon and Schuster.

Weiss, L. and Hobson, J. M. (1995). *States and Economic Development: A Comparative Historical Analysis*. Cambridge: Polity Press.

Wesson, R. G. (1967). *The Imperial Order*. Berkeley, CA: University of California Press.

Wesson, R. G. (1978). *State Systems: International Pluralism, Politics, and Culture*. New York: Free Press.

Wigmore, J. H. (1928). *A Panorama of World's Legal Systems*. Volumes I, II and III, St. Paul, MN: West Publishing.

Wilkinson, D. (2004). The Power Configuration Sequence of the Central World System, 1500 – 700 BC, *Journal of World-Systems Research X*, 3, 655-720.

Wittfogel, K. A. (1957). *Oriental Despotism: A Comparative Study of Total Power*. New Haven, CN: Yale University Press.

Wittman, D. (1991). Nation and states mergers and acquisitions dissolutions and declines. *American Economic Review 81*, 126-129.

Wolpert, S. (2004). *A New History of India*.7th Ed.. Oxford: Oxford University Press.

Wong, K. Y. (1995). *International Trade in Goods and Factor Mobility*. Cambridge, MA: MIT Press.

Zane, J. M. (1927). *The Story of Law*. 2nd Ed. Indianapolis, IN: Liberty Fund.

Index

A

Abbasid 16, 95, 96, 101, 102, 103, 104, 122, 124, 131, 141, 146
Abbasid Caliphate 96, 101, 103, 104, 122, 131, 141, 146
Abbasid Golden Age 95, 102, 104, 105, 124
Absolutism 149
Academy of Gundishapur 85
Achaemenid 34, 57, 60, 61, 62, 63, 64, 66, 70, 79, 82, 83, 142
Acheh 39
Acropolis 67
adjustable centerboard keel 117
Aegean Sea 21, 140
Afghanistan 82, 84, 89, 92, 144
Africa 9, 29, 36, 39, 40, 41, 100, 102, 105, 107, 108, 141, 146, 153, 158, 160
African 29, 98, 104, 107, 108
Afshar 35, 142, 144
Aghlabids 101
Ahura Mazda 63
Akkadian 33, 48, 49, 56
Alexander the Great 63, 66, 68, 71, 72, 79
Alexandria 68, 107
Algeria 69, 101
al-Ghazali 103
Alhazen 161
al-Ma'mun 104
Alp-Arslan 102
alphabetical 50

Alps 109
America 9, 36, 39, 41, 148, 153, 165
American 13, 40, 41, 166
Anatolia 49, 54, 63, 92, 140
Andras Empire 88
An Lu Shan 113, 114, 122
Anti-Taurus 48
Arab 82, 103
Arabian 9, 48, 63, 84, 97, 98, 100, 101, 102, 103, 105, 110, 113, 122, 123, 155
Arabs 35, 98, 101
Aral Sea 63, 89, 102, 144
Aristophanes 67
Aristotle 67, 68
Armenia 82, 84
Arsacid 82
Arthashastra 73, 88
Aryan 45, 54, 70, 71
Ashoka 73, 88
Ashtadyayi 90
Asiatic 97, 153
Assurbanipal 56
Assyrian 33, 34, 49, 54, 55, 56, 57, 60, 61, 62, 63, 64
Atenism 57
Athenian Empire 70
Athens 64, 66, 67, 68, 70
Australia 40, 41, 153
Austria 27, 141, 151, 154
Ava 39
Avicenna 104
axial age 9
Axial Period 59, 62

Legalist 75, 76, 112
Levant 54, 100, 107, 141
Liang Dynasty 111
Liao 114, 115
Liao Kingdom 114, 115
Libya 69
Light cavalry 34
light infantry 34, 63, 72, 89
LiKui 76
Lilybaeum 69
Liu Bei 111
logicians 76
London 152
Lord Shang 76
Louis XIV 150, 151, 157
Lu Bu Wei 43, 77

M

Macedonia 66, 67, 80
Macedonian 65, 68, 80, 88
madrasahs 103
Magadha 70, 88
Mahabharata 90
Mahabhashya 90
mahajanapada 70
Mahavira 72
Majapahit 39, 91
major cultures 8, 9, 39, 84, 85, 91, 92, 95, 97, 100, 119, 122, 123, 124, 125, 128, 131, 145, 146, 147, 149, 163, 164
Malaya 31, 91, 106
Malay Peninsula 39
Malaysia 90, 165
Maldives 90, 106
Malikshah 102, 103
Malukas 39
Mamluk 102, 145
Manava Dharmashastra 90
Manchuria 100
Manchurian 77, 114, 131, 132, 137, 141
Mandalay 39
Marco Polo 9
marginal effect of relative capability 22, 23, 24, 84
Maria Theresa 154

mass factor 19, 20, 22, 23, 24, 25, 26, 28, 29, 30, 31, 32, 34, 37, 39, 41, 48, 49, 55, 57, 60, 61, 63, 67, 69, 71, 72, 75, 81, 84, 86, 91, 92, 95, 96, 99, 106, 108, 109, 110, 111, 112, 122, 124, 127, 128, 130, 131, 132, 136, 138, 139, 140, 141, 142, 148, 149, 150, 151, 163, 164, 165, 166
Mataram 39
Mathura school of art 89
Matteo Ricci 134
Mauryan 27, 71, 73, 88, 89, 92
McNeill 15, 54, 57, 91, 97, 98, 99, 100, 104, 107, 110, 116, 118, 119, 149, 150, 157, 158, 159, 160, 164
Median 82
medieval 9, 13, 23, 31, 36, 37, 72, 82, 83, 85, 89, 92, 95, 97, 99, 100, 101, 102, 103, 104, 105, 106, 107, 108, 109, 110, 111, 112, 113, 119, 120, 121, 122, 123, 124, 127, 128, 129, 130, 132, 136, 138, 139, 140, 141, 142, 144, 145, 146, 148, 149, 150, 161, 164
Mediterranean 21, 22, 34, 35, 51, 61, 62, 64, 65, 66, 67, 69, 78, 80, 81, 82, 84, 88, 93, 95, 98, 100, 102, 107, 108, 110, 123, 140, 141, 146
Mediterranean Sea 81, 98, 140
Meiji Restoration 28, 165
Melaka 39
mercantilism 43, 151, 152
Merv 142
Mesoamerica 41
Mesopotamia 9, 33, 45, 46, 47, 48, 49, 50, 51, 52, 53, 54, 55, 57, 58, 60, 62, 63, 82, 92, 102, 140, 141, 147
Middle East 35, 36, 54, 56, 62, 64, 71, 80, 82, 83, 84, 93, 99, 100, 102, 105, 108, 109, 116, 121, 122, 123, 124, 128, 129, 131, 136, 140, 142, 155, 156
Middle Kingdom 47, 52, 131
military decisiveness 19, 41, 80, 128
military-industrial complex 158
military revolution 15, 22, 32, 34, 35, 66, 67, 81, 93, 97, 99, 111, 121, 127,